The Sword's Other Edge

This book is the first work to build a conceptual framework describing how the pursuit of military effectiveness can present military and political trade-offs, such as undermining political support for the war, creating new security threats, and how seeking to improve effectiveness in one aspect can reduce effectiveness in other aspects. Here are new ideas about military effectiveness, covering topics such as military robotics, nuclear weapons, insurgency, war finance, and public opinion. The study applies these ideas to World War II, the Korean War, the Vietnam War, and the 1973 October War as well as to ongoing conflicts and public policy debates, such as the war on terror, drone strikes, ISIS, Russian aggression against Ukraine, US–Chinese–Russian nuclear competitions, and the Philippines insurgency, among others. Both scholarly and policy-oriented readers will gather new insights into the political dimensions of military power and the complexities of trying to grow military power.

DAN REITER is the Samuel Candler Dobbs Professor of Political Science at Emory University. He is the award-winning author of three books, *Crucible of Beliefs: Learning, Alliances and World Wars*, *Democracies at War* (with Allan C. Stam), and *How Wars End*, as well as dozens of articles about the causes, prosecution, and termination of war, alliances, domestic politics and international relations, nuclear weapons, terrorism, and other topics. He is a recipient of the Karl Deutsch Award, given annually to the leading scholar of international relations under age forty or within ten years of having received a PhD.

The Sword's Other Edge

Trade-offs in the Pursuit of Military Effectiveness

Edited by

DAN REITER

Emory University

CAMBRIDGE
UNIVERSITY PRESS

CAMBRIDGE
UNIVERSITY PRESS

University Printing House, Cambridge CB2 8BS, United Kingdom

One Liberty Plaza, 20th Floor, New York, NY 10006, USA

477 Williamstown Road, Port Melbourne, VIC 3207, Australia

314–321, 3rd Floor, Plot 3, Splendor Forum, Jasola District Centre, New Delhi – 110025, India

79 Anson Road, #06-04/06, Singapore 079906

Cambridge University Press is part of the University of Cambridge.

It furthers the University's mission by disseminating knowledge in the pursuit of education, learning and research at the highest international levels of excellence.

www.cambridge.org
Information on this title: www.cambridge.org/9781108416726
DOI: 10.1017/9781108241786

First published 2017

Printed in the United States of America by Sheridan Books, Inc.

A catalogue record for this publication is available from the British Library.

Library of Congress Cataloging-in-Publication Data
Names: Reiter, Dan, 1967– editor of compilation.
Title: The sword's other edge : tradeoffs in the pursuit of military effectiveness / edited by Dan Reiter, Emory University.
Other titles: Tradeoffs in the pursuit of military effectiveness
Description: New York : Cambridge University Press, [2017] | Includes bibliographical references and index.
Identifiers: LCCN 2017023082 | ISBN 9781108416726 (alk. paper)
Subjects: LCSH: Military readiness. | Military policy – Political aspects. | Military art and science – Case studies.
Classification: LCC UA10 .S867 2017 | DDC 355/.0332–dc23
LC record available at https://lccn.loc.gov/2017023082

ISBN 978-1-108-41672-6 Hardback
ISBN 978-1-108-40413-6 Paperback

Contents

Figures

Tables

Contributors

Dr. Filippo Andreatta teaches International Politics and Strategic Studies at the University of Bologna and is Director of IP LAB in Trento.

Dr. Emanuele Castelli is Assistant Professor in Political Science at the University of Parma.

Dr. Joseph Felter is a fellow at the Hoover Institute at Stanford University.

Dr. Jeffrey A. Friedman is Assistant Professor of Government at Dartmouth College.

Dr. Michael C. Horowitz is Professor of Political Science and Associate Director of Perry World House at the University of Pennsylvania.

Dr. Jason Lyall is Associate Professor of Political Science and Director of the Political Violence FieldLab at Yale University.

Dr. Dan Reiter is the Samuel Candler Dobbs Professor of Political Science at Emory University.

Dr. Caitlin Talmadge is Assistant Professor of Political Science and International Affairs at the George Washington University.

Dr. Lorenzo Zambernardi is Assistant Professor in the Department of Political and Social Sciences at the University of Bologna.

Dr. Rosella Cappella Zielinski is Assistant Professor of Political Science at Boston University.

Acknowledgments

This project originated within the activities of the Research Center on International Politics and Conflict Resolution, created in 2012 by the Bruno Kessler Foundation (FBK) in Trento, Italy. Earlier drafts of most of the chapters of this volume were originally presented at the Military Effectiveness Research Seminar in Trento, June 23–25, 2015. We would like to acknowledge the generous financial support FBK provided for the seminar. We also are deeply grateful for the many thoughtful comments provided at the seminar by Joseph Grieco, Tyson Chatagnier, Ted Hopf, and Pascal Vennesson. Our great thanks as well to Moira Osti for all her able logistical efforts.

1 *Confronting Trade-offs in the Pursuit of Military Effectiveness*

DAN REITER

Since antiquity, perhaps the most commonly studied question of international politics has been, Who wins wars? Very regrettably, violent conflict in the international system has endured into the twenty-first century, notwithstanding the "End of History" hope that the close of the Cold War might usher in a new era of permanent peace. Many wars between states have been waged since 1990, and even more conflicts within states have erupted and raged. Millions have died in violent, political conflict since then, and there is no end in sight to this curse on humanity.

This grim legacy means that twenty-first-century scholars and policy makers still must think about how both to avoid conflict and to use force effectively in the event of conflict. Even peace-loving leaders need to think about using force effectively, to deter and if necessary fend off violent attacks and in pursuit of humanitarian missions. Leaders must think both about how to engage effectively in long-standing forms of conflict, such as conventional and insurgent wars, and about how to engage in new forms of conflict, such as cyberwarfare and drone strikes.

This book endeavors to help scholars and policy makers better understand the pursuit of military effectiveness, past, present, and future, making two central contributions. First, it pushes the study of military effectiveness in new directions, exploring elements of military effectiveness that are just emerging in the twenty-first century and applying new ideas to the study of enduring questions of military effectiveness. Though this is a scholarly area that has received substantial attention, especially in recent years, the contributions in this book demonstrate that there is yet much to be learned.

Second, the book presents and develops an essential and often underappreciated proposition: *the pursuit of military effectiveness often incurs important trade-offs*. Most phenomena or policies thought to affect military effectiveness do not have simple, limited consequences, either for a belligerent's likelihood of success in combat or for other values the belligerent might hold dear. The contributions in this volume each discuss a particular element of military effectiveness within the context of exploring how the pursuit of that aspect of effectiveness risks incurring trade-offs. The trade-offs discussed in these chapters fall into three categories: political support, security threats, and war fighting. Failure to recognize these trade-offs encourages a reductionist, incomplete view of military effectiveness that falls short of a complete understanding of war and undermines quality decision making.

This chapter serves to frame the rest of the book. It first discusses the concept of military effectiveness, presenting a new definition. It then lays out the three categories of trade-offs. Last, it presents a road map of the chapters in the rest of the volume.

Thinking about Military Effectiveness

Scholars, political leaders, and generals back to antiquity have considered what factors might help militaries win wars, including ancient Greek historians such as Herodotus and Thucydides as well as Chinese thinkers such as Sun Tzu. Machiavelli paid great attention to mastery of the art of war as part of the art of statecraft. Enlightenment thinkers and political leaders such as John Locke, Thomas Jefferson, John Quincy Adams, Alexis de Tocqueville, and Thomas Paine applied their broader political ideas to narrower military effectiveness questions, for instance, examining factors that determine how effectively a society can muster resources for war and motivate its soldiers to fight and die in battle.[1]

Observers past and present have focused not only on how belligerents win wars but also on how belligerents and their militaries accomplish the smaller, instrumental tasks that accumulate to eventual victory in war. Focusing on the individual tasks within war rather than just observing who wins and loses wars enables avoiding the assumption that war winners performed all war-related tasks well and war losers performed all war-related tasks poorly. History is, of course, littered with examples of militaries that performed many war-related tasks well but still lost the war, such as Germany in World War II, or performed

[1] Reiter and Stam (2002, esp. Chapter 3); Gilbert (1986).

many war-related tasks poorly but managed to win the war, such as the Soviet Union in the Winter War.

There is space here to survey very briefly only recent political science scholarship on military effectiveness, a limited selection of the vast amount of social scientific and historical work on military effectiveness. In general, modern political scientists have examined a range of phenomena associated with the successful accomplishment of tasks that contribute to victory in war. A mainstream line of thinking has emphasized the importance of military–industrial power for winning wars and in turn explored whether factors such as domestic politics or having a central bank affect the ability of belligerents to acquire the economic resources needed to field powerful militaries.[2] Other work has examined the sophistication and the size of belligerents' armed forces, examining factors that make a military more or less likely to adopt (effectively) a particular strategic or technological innovation, and in turn other studies have examined whether higher levels of military technology affect conflict outcomes.[3] Other studies have argued that it is not only the forces being deployed in combat but also the plans for the deployment of forces in combat that strongly affect effectiveness. Military strategy, doctrine, and force employment are all seen as determinants of operational success in conflicts such as conventional wars, aerial bombing campaigns, and nonconventional conflicts such as counterinsurgency and counterterrorism campaigns (and scholars have also looked at factors, such as domestic politics and combat outcomes, that shape how states choose their strategies, doctrines, and force employment strategies).[4] Several works have taken Napoleon Bonaparte's dictum that "the moral is to the physical as three to one"

[2] Organski and Kugler (1980); Desch (2008); Beckley (2010); Bueno de Mesquita et al. (2003); Schultz and Weingast (2003); Lake (1992); Reiter and Stam (2002, Chapter 5); Cappella Zielinski (2016); Poast (2015); Shea (2014).

[3] Evangelista (1988); Rosen (1991); Gartner (1997); Horowitz (2010); Biddle and Zirkle (1996); Goldman and Eliason (2003); Biddle and Long (2004); Resende-Santos (2007); Lyall and Wilson (2009); Biddle (2004); Levi (1997).

[4] Biddle (2004); Avant (1994); Kier (1997); Arreguín-Toft (2005); Gartner (1997); Mearsheimer (1983); Stam (1996); Pape (1996); Horowitz and Reiter (2001); Krepinevich (1986); Johnston (2012); Johnston and Sarbahi (forthcoming); Peic (2014); Lyall (2009, 2010); Price (2012); Long (2014); Jordan (2009); Reiter and Meek (1999); Snyder (1984); Merom (2003); Brooks (2008); Grauer and Horowitz (2012). For a discussion on several aspects of military effectiveness, see the contributions in Brooks and Stanley (2007).

to suggest the importance of troop morale for affecting military effectiveness, exploring factors affecting soldiers' motivations to fight and die, such as nationalism, regime type, small group bonding, and social cleavages.[5]

Further studies have explored a variety of other sources of military effectiveness, such as military leadership quality, civil–military relations, human capital factors such as education, and wartime coalitions.[6] Some have also developed more subtle arguments, such as that the ability to win wars is strongly determined by the decision to start wars and that states that are smarter about starting wars, such as democracies, are more likely to win.[7] Others have proposed that the likelihood of victory is related to the nature of the stakes the states are fighting over.[8]

Though debates continue to rage over the sources of military effectiveness, and the sources of military effectiveness vary across different modes of conflict, one overall pattern is that nonmaterial sources of military effectiveness, such as military strategy, domestic political factors, nationalism, and military leadership, are likely at least as important as military factors, such as the balance of forces or sophistication of weaponry.

This body of work has examined factors that affect the abilities of belligerents to accomplish tasks that contribute to victory. We define military effectiveness in line with this theme, conceiving military effectiveness as *the degree to which militaries can accomplish at acceptable costs the goals assigned to them by political leaders*. We make two observations about this definition. First, within the political ends–military means chain of grand strategy, our definition focuses on military means, leaving aside how leaders conceptualize political ends. Second, though the definition incorporates efficiency, it does not equate effectiveness with efficiency. In this regard, our definition deviates slightly from the efficiency emphasis of the widely used definition of

[5] Shils and Janowitz (1948); Levi (1997); Peled (1998); Reiter and Stam (2002, Chapter 3); Castillo (2014); Rosen (1996); Reiter (2007); Posen (1993); Lyall (2016). For critique of the small group thesis, see King (2016).

[6] Rotte and Schmidt (2003); Reiter and Wagstaff (forthcoming); Quinlivan (1999); Brooks (2008); Pilster and Böhmelt (2011, 2012); Choi (2003, 2012); Lake (1992); Biddle and Long (2004); Talmadge (2015).

[7] Bueno de Mesquita (1981); Reiter and Stam (2002); Weeks (2014). See also Desch (2008); Brown et al. (2011).

[8] Sullivan (2012).

military effectiveness crafted by Allan Millett and his colleagues: "the process by which armed forces convert resources into fighting power."[9] Though efficiency, the accomplishment of military tasks at lower costs, is an important component of effectiveness, the two terms should not be viewed as equal. For example, some belligerents deliberately adopt strategies and tactics they know to be costly, and therefore perhaps inefficient, because they view such approaches as necessary to achieving their desired goals and are willing to pay higher costs to accomplish these goals. Examples of inefficient but effective approaches include certain Soviet ground warfare tactics in World War II, including the use of blocking units described in Chapter 4, authored by Jason Lyall, in this volume, and human wave tactics employed by China in the Korean War. One might also view American economic provision of weapons and other logistical support during World War II as both inefficient, in the sense that there was a tremendous amount of waste, and effective, in the sense that these efforts were critically important for Allied victory.[10]

Trade-offs in the Pursuit of Military Effectiveness

Some existing scholarship on military effectiveness takes a narrow approach, evaluating the connection between a single causal phenomenon and a single element of military effectiveness, such as whether nationalism affects soldiers' willingness to fight. Policy makers also sometimes take narrow views of proposals to advance military effectiveness, thinking about specific effects of a particular weapon system or strategy and failing to consider the variety of possible consequences that might ensue. For example, the 1961 American decision to use herbicides, such as Agent Orange, in the insurgency conflict in South Vietnam apparently failed to consider a number of possible drawbacks, such as degradation of agricultural production for non–Viet Cong South Vietnamese civilians and negative health consequences for South Vietnamese civilians and US forces.[11]

One classic observation about unexamined consequences of fighting wars is the concern that war, even victory in war, may impose a

[9] Millett et al. (1988, 2). On defining military effectiveness, see also Brooks and Stanley (2007, 9); Talmadge (2015, 5); Rosen (1995, 6).

[10] Atkinson (2007); Overy (1995).

[11] Buckingham (1983).

sacrifice of national values, including human and economic costs, that exceeds the stakes of the conflict. A military victory that incurs unacceptably high costs has been called a Pyrrhic victory, so named from the experience of the Greek king Pyrrhus of Epirus. Pyrrhus's army defeated Roman forces at Heraclea and Asculum around 279 BCE at the expense of unacceptably high Epirean battle fatalities. Plutarch records that after being told of a battlefield victory by one of his commanders, Pyrrhus replied, "If we are victorious in one more battle with the Romans, we shall be utterly ruined."[12] This core insight was famously reframed by Carl von Clausewitz in his nineteenth-century masterpiece *On War*, where he stated famously that war is politics "carried on with other means,"[13] the means of war serving political ends. And certainly a number of modern conflicts have been critiqued as being waged at costs dwarfing the issues at stake. The most horrifying war of the modern era may have been the 1860s Lopez War, in which Paraguay lost some 50 percent of its prewar population.[14] Britain spent a staggering 2 percent of its annual gross domestic product to wage a military expedition in 1867–1868 in Abyssinia to rescue a mere seven European hostages.[15] Bernard Brodie critiqued European decision making before and during World War I, proposing that European leaders lost sight of the political motivations for fighting.[16] Many have made a similar point about the Vietnam War, sometimes pointing to the 1968 statement of an American major that "it became necessary to destroy the town in order to save it" as reflective of the loss of American political perspective during the war. More colorfully, the absurdity of the decades-long India–Pakistan conflict over the Siachen glacier has been captured by the comparison of the conflict with "a struggle of two bald men over a comb."[17]

[12] http://penelope.uchicago.edu/Thayer/E/Roman/Texts/Plutarch/Lives/Pyrrhus.html#21.

[13] Clausewitz (1976, 87) [14] Clodfelter (2008, 346).

[15] Caverley (2014, 156–157). One part of what Millett et al. (1988) describe as "strategic effectiveness," linking military operations within broader foreign policy goals, might be seen as allowing for the inclusion of political costs. But accounting for foreign policy goals in this way is beyond their general definition of military effectiveness, the conversion of resources into fighting power. The question of how well military operations fit within broader foreign policy goals is more generally framed within the purview of grand strategy rather than as part of military effectiveness.

[16] Brodie (1973). [17] www.bbc.com/news/world-asia-india-26967340.

Beyond the "Pyrrhic victory" claim, scholars and policy makers have sometimes considered the possibility that the pursuit of military effectiveness risks other kinds of trade-offs. However, past discussions of these trade-offs have occurred mostly in isolation from each other, and scholars have not yet taken the next step of recognizing the ubiquity and similarities of these trade-offs. This chapter, and the book more generally, advances our understanding of military effectiveness by taking this next step, proposing that trade-offs are common to the pursuit of military effectiveness and that there are important common elements that tie these trade-offs together into categories. Specifically, this chapter builds a conceptual framework for three main types of trade-offs: *political support*, *security threats*, and *war fighting*. The chapter describes the trade-offs themselves as well as past scholarship and historical episodes that describe these trade-offs. Each of the chapters that follow describes in greater theoretical and empirical detail the presence of one or more of these trade-offs within a single area of military effectiveness.

Political Support

If, as Clausewitz wrote, war is politics by other means, then necessarily wars are fought in political contexts. One context is domestic politics. The pursuit of military effectiveness might undermine public support for war by increasing friendly military casualties, friendly civilian casualties, and/or the financial costs of war. There has been substantial scholarly discussion of the first category, how the escalation of friendly military casualties risks undermining public support for war, and it is the central theme of Chapter 2, on force protection, by Emanuele Castelli and Lorenzo Zambernardi. A substantial body of scholarship has documented that increases in friendly military casualties can reduce public support for war, for the simple reason that as casualties climb, the public becomes less likely to view the war as worthwhile. There are nuances to this relationship, as, for example, if the stakes are high enough, then higher friendly casualties might not reduce support, and an individual's support for war is more likely to be affected negatively by casualties if the people who died had lived in the individual's geographic region or are part of the individual's social network. Some also speculate that other factors, such as elite rhetoric or public perceptions on whether the war is being won, have greater effects on public support than casualty rates. But overall, the supporting empirical evidence for the casualties–public support proposition is impressive.[18]

Increased friendly *civilian* casualties might also undermine public support for war. This question received great interest in the 1920s and 1930s, when theorists speculated that strategic bombing of cities and the accumulation of civilian casualties in the next war would quickly push publics to demand their governments to end the war. Any state whose public was vulnerable to strategic bombing would be unable to fight long wars. Strikingly, however, the historical record indicates that bombing civilians failed to increase public opposition to war and in some case may have had the opposite effect.[19] Occasionally, belligerents take the related tack of seeking to undermine an adversary's public's support for war by using the adversary's civilians as human shields, the idea being that the adversary's public will shy away from war because those captured civilians might be jeopardized. As Saddam Hussein learned in fall 1990, this strategy is ineffective.[20]

Conflict scholars have drawn another connection between friendly civilian deaths and public support for war with regard to counterinsurgency (COIN). Mainstream COIN theory recognizes that counterinsurgent attacks that kill civilians undermine public support for COIN forces, making victory more difficult. Others also speculate that insurgent attacks on civilians undermine public support for insurgents. Indeed, the escalating brutality of al-Qaeda against Iraqi civilians during the postinvasion Iraqi insurgency in the 2000s may have been one reason why the tide of the war shifted in favor of US COIN forces in late 2007.[21] American air strikes on South Vietnamese targets during the Vietnam War increased local support for the Viet Cong.[22] Some empirical work has found different relationships, for example, one study found that Russian artillery strikes on Chechen villages in the Second Chechen War had the effect of reducing Chechen insurgent violence.[23]

[18] Mueller (1973); Gartner et al. (1997); Gartner and Segura (1998); Gartner (2008a, 2008b); Gartner and Segura (2008); Berinsky (2009); Gelpi et al. (2009).

[19] Pape (1996); Horowitz and Reiter (2001).

[20] www.cia.gov/library/reports/general-reports-1/iraq_human_shields.

[21] Ricks (2009a).

[22] Kocher et al. (2012). See also Kalyvas (2006).

[23] Lyall (2009).

A third avenue by which the pursuit of military effectiveness might reduce public support for war is if such pursuits are financially costly. War costs money, and citizens may not be willing to sacrifice butter to acquire guns. There is a growing body of literature on the connection between financial expenditures and support for war, and it is the main topic of Chapter 3, authored by Rosella Cappella Zielinski.[24]

Beyond the intrinsic costs of greater bloodshed and the diversion of economic resources, the trade-off of eroded public support for war has other consequences, including potentially affecting a belligerent's prosecution of its war, forcing a belligerent to exit a war short of (decisive) victory, and undermining the belligerent's leadership's hold on political power. Regarding a belligerent's prosecution of war, eroded public support or the prospect of eroding public support can have several effects on how a war is fought. An excessive focus on minimizing friendly casualties, or force protection, can undermine aggressiveness, slowing the ability of an army to capture territory. In the American Civil War, for example, President Abraham Lincoln tabbed Union general George McClellan as having a case of the "slows," as

> McClellan moved his forces with caution, often out of concern for risking higher casualties. Only when McClellan was replaced by more aggressive generals such as William T. Sherman, Ulysses Grant, and Philip Sheridan who were more willing to accept Union casualties, was the Union finally able to break the back of the Confederate Army and save the American republic.[25]

Force protection concerns can encourage other military strategies, including capital-intensive strategies that emphasize firepower over labor-intensive strategies that risk troops.[26] Others have proposed that states sensitive to friendly casualties may adopt military strategies that may reduce friendly casualties at the expense of increasing foreign casualties, including foreign civilian casualties.[27] The most extreme example of this last dynamic, perhaps, is the 1945 American decision to drop atomic bombs on the Japanese cities of Hiroshima and Nagasaki in hopes of ending the war before a very bloody American amphibious invasion of the Japanese home islands might become necessary.[28]

The other side of the force protection trade-off is that leaders sometimes recognize that they must accept friendly casualties to accomplish

[24] See Flores-Macías and Kreps (2013, forthcoming); Kriner et al. (2015).
[25] McPherson (1988). [26] Caverley (2014). [27] Downes (2008).
[28] Frank (1999).

important operational goals. The World War II American general George Patton was ruthlessly willing to sacrifice his own troops in pursuit of operational objectives, such as speed of advance. During the 1943 Sicily campaign in World War II, Patton remarked that if his forces could reach Messina a single day earlier if it meant "losing additional men," then he would be willing to "lose them."[29] Perhaps accordingly, Patton's troops sometimes cynically interpreted his nickname of "Old Blood and Guts" as "our blood, his guts."

War-fighting strategies aside, declining public support may constrain a belligerent's access to human and economic resources. At the limit, publics may riot in protest of conscription, as occurred in the Union during the American Civil War.[30] The specter of opposition may push leaders to seek an end to the war rather than fighting with insufficient resources or risking political upheaval. Following the January 1968 Tet offensive in the Vietnam War, the Johnson administration rejected the military-recommended option of escalation of American involvement, in part because doing so would increase American casualties and impose substantial strains on the American economy, both of which would further undermine public opinion. Democracies in particular face these possibilities of escalating human and financial[31] costs pushing elected leaders to exit wars faster. Indeed, General George C. Marshall once observed, "A democracy cannot fight a seven years' war."[32]

The pressures presented by this public support trade-off toward, for example, fighting shorter wars or minimizing civilian casualties may have positive or negative effects on military effectiveness, as described in Castelli and Zambernardi's Chapter 2. For example, a maneuver-based strategy, chosen to keep the war short and low cost, may also make victory more likely.[33] Jonathan Caverley proposed that American military strategy in Vietnam was ineffective in part because the imperative toward force protection encouraged an emphasis on firepower over manpower, an approach ultimately ineffective in fighting insurgencies.[34]

[29] Quoted in Atkinson (2007, 143).

[30] On conscription and public support, see Vasquez (2005); Horowitz, Simpson, and Stam (2011); James and Choi (2005). On the Civil War, see McPherson (1988).

[31] Hallek (1998, 509); Herring (1996, 222–223).

[32] Reiter and Stam (2002, Chapter 7).

[33] Reiter and Stam (2002).

[34] Caverley (2014).

This cluster of propositions, of course, presumes that dissatisfied publics can turn their leadership out of power. There is a substantial scholarship evaluating this proposition. A long-standing view, dating back to Immanuel Kant in the eighteenth century, is that publics unhappy with war can more easily turn out of power elected leaders than authoritarian leaders. This insight in turn served as the foundation for the modern democratic peace proposition.[35] Especially more recently, scholars have considered variations of this idea. For example, Jessica Weeks proposed that some authoritarian leaders, namely, those who run party "machine" regimes, can be turned out of power by dissatisfied elites, and as a result, their foreign policy behavior converges with that of democracies.[36] The empirical evidence on the connection between war outcomes and leader tenure, accounting for variance in political institutions, is surprisingly nuanced. For example, there are varying effects on the likelihood of leaders exiting from power through regular means, such as losing elections, as compared with exiting from power through irregular means, such as coups d'état. Furthermore, leaders who start wars may be more likely to suffer the domestic political consequences of defeat, as compared with leaders who end a war in defeat but took power after the war had already begun.[37]

Security Threats

A second major category of trade-offs concerns security threats. Attempts to improve military effectiveness can create security threats. Improving effectiveness during peacetime can spark hostility, fueling arms races. Expanding one's military or mobilizing in the context of a crisis can fuel security dilemmas and attract preventive or preemptive attacks.[38] Some attempts by a belligerent to improve military effectiveness against one adversary can have the effect of creating new security threats by provoking conflicts with other adversaries or by undermining the belligerent's ability to fight effectively against other adversaries. Regarding provoking new conflicts, some attempts to improve effectiveness risk encroaching on the interests or territory of neutrals, causing those affected neutrals to declare war. World War I Germany made perhaps the two most famous efforts to improve

[35] Russett (1993). [36] Weeks (2014).
[37] Chiozza and Goemans (2011); Croco (2015). [38] Jervis (1978).

military effectiveness that created new security threats by incurring conflict with states that otherwise might have remained neutral. In August 1914, Germany's invasion of Belgium as part of its strategy to conquer France ensured Britain's entry into the war on the side of Russia and France.[39] British belligerence in turn meant British troops fighting against German forces in the West, and a maritime blockade of Germany imposed by the Royal Navy. In early 1917, Germany decided to enhance the effectiveness of its submarine blockade of Britain by announcing its intention to attack all neutral shipping in the Atlantic. This decision helped push the US to declare war on Germany in April 1917, and American forces deployed to the Western Front in spring 1918 were critical in helping block the last German offensives, and force German collapse.[40] Caitlin Talmadge's chapter in this volume addresses a related form of security threat trade-off, how conventional military operations might escalate to nuclear exchanges.

Fear of such trade-offs in turn sometimes shapes how belligerents fight wars. States may eschew some wartime efforts that could improve effectiveness and increase the chance of victory over the adversary, if those efforts risk bringing neutrals into the war on the side of the adversary. In the Korean War, fear of drawing in the Soviet Union discouraged the US from launching bombing attacks on Chinese territory.[41] Similarly, American bombing and troop escalation decisions during the Vietnam War were constrained by a strong desire to avoid attracting Chinese participation.[42]

Past wars have demonstrated an array of different connections between the pursuit of military effectiveness and the actions of neutral states. On the morning of October 6, 1973, the Israeli government received warning of likely Arab attack that day, but declined the militarily attractive option of launching a preemptive attack, as it had done in June 1967 to commence the Six-Day War, out of fear that doing so would jeopardize American support.[43] During the American Civil War, President Abraham Lincoln's Emancipation Proclamation highlighted the role of slavery in the war, in turn helping discourage France and Britain from entering the war on the side of the Confederacy.[44] During the 1991 Gulf War, Saddam Hussein's Iraq launched missile attacks on Israeli cities, hoping that doing so would reframe the war as a conflict

[39] Tuchman (1962). [40] Goemans (2000). [41] Reiter (2009, Chapter 5).
[42] Khong (1992). [43] Reiter (1995). [44] Reiter (2009).

between Arabs and Israel, in turn encouraging Arab states to oppose the UN Coalition. This gambit failed.

The use of weapons of mass destruction (WMD) can also affect security threats by influencing the decisions of neutrals, in part because of the normative dimensions of their use.[45] The 2013 Syrian government use of chemical weapons in its civil war nearly invited American air strikes, averted at the last moment when Syria brokered a deal to destroy its chemical arsenal.[46] Four years later, a new round of Syrian chemical weapons attacks cause the US to launch cruise missile strikes on a Syrian air base. Anticipation of adverse neutral actions sometimes shapes the WMD decisions of states. Thomas Risse-Kappen argued that the main reason the US did not use nuclear weapons during the Korean War was European diplomatic opposition.[47] Some speculate that the atomic bombing of Hiroshima in August 1945 may have hastened the Soviet decision to declare war on Japan.[48]

Aside from affecting the decisions of neutrals, the pursuit of military effectiveness can also incur security threat trade-offs by drawing off resources from other military and foreign policy priorities, as Rosella Cappella Zielinski and Joseph Felter describe in their chapters. This point has been made in the context of evaluating the decision to initiate wars. For example, a common critique of the 2003 invasion of Iraq was that it diverted US military resources from Afghanistan that were necessary to continue to defeat Taliban and al-Qaeda forces.[49] The German declaration of war on the US in December 1941 forced Germany to divert resources away from its titanic, ongoing war with the Soviet Union. The American decision in June 1965 to deploy five hundred thousand troops to the Vietnam War stripped away forces that otherwise could have been used to protect other American allies.[50] Furthermore, structuring one's military to fight effectively against some kinds of threats might reduce its ability to fight against other kinds of threats. Stanislav Andreski posited that Latin American dictators tended not to fight each other because their militaries were structured to crush internal foes.[51] Conversely, Andrew Krepinevich proposed that in the 1960s the American Army was structured organizationally and

45 Tannenwald (1999).
46 On norms against using chemical weapons, see Price (1997).
47 Risse-Kappen (1995). 48 Hasegawa (2005, 187). 49 Ricks (2006).
50 Kaplan (2006). 51 Andreski (1980).

doctrinally to fight massive conventional conflicts, an orientation that in turn impeded its ability to combat the Viet Cong insurgency.[52]

War Fighting

A third type of trade-off concerns war fighting against the belligerent in question. That is, a belligerent's attempt to improve its military effectiveness against a particular adversary may in other ways reduce the belligerent's military effectiveness against that same adversary. Some might frame such a boomerang effect in the context of thinking about different levels of military effectiveness, perhaps using the Millett et al. categories of political, strategic, operational, and tactical levels of military effectiveness.[53] Though these are helpful categories, the discussion below is not constrained to considerations of trade-offs among or within these levels.

There are several examples of this kind of war-fighting trade-off, beyond those discussed in this volume in the Michael Horowitz and Jason Lyall chapters in particular. One concerns efforts to motivate troops through calls to nationalism. The attractiveness of such efforts from a tactical military effectiveness perspective is straightforward. However, some of the measures a belligerent could take to advance these immediate goals might have a variety of negative consequences. Amping up soldiers' nationalism might encourage the dehumanization of the opponent's civilians and soldiers, making war atrocities more likely, as scholars of the World War II Soviet and Pacific theaters have described.[54] Greater incidence of war crimes could undermine war-fighting effectiveness. If an army mistreats surrendering enemy soldiers, this can make enemy soldiers less willing to surrender and more likely to fight to the death.[55]

Counterinsurgency in particular presents a number of possible war-fighting trade-offs. As discussed above, killing civilians can undermine popular support and make victory more difficult. Some have proposed that the use of mechanized transport can reduce effectiveness by impeding the collection of intelligence.[56] Development aid can under

[52] Krepinevich (1986). [53] Millett et al (1988).
[54] Bartov (1992); Dower (1986).
[55] Reiter and Stam (2002, Chapter 3); Reiter (2007); Grauer (2014).
[56] Lyall and Wilson (2009); see also Peic (2014).

some conditions increase insurgent violence.[57] There is ongoing debate about the utility of torture as a counterinsurgency tool. Its effectiveness as interrogation tool notwithstanding, it presents the worrisome trade-off of fueling sympathy for the terrorist or insurgent group, likely encouraging more individuals to join the group.[58] In 2006, Secretary of State Donald Rumsfeld conceded the possibility of this kind of trade-off, noting that regarding the Iraq War, "Are more terrorists being created in the world? We don't know."[59]

Dangers of Ignoring or Misunderstanding Trade-offs

Ignoring possible trade-offs can have deleterious consequences for both scholars and policy makers. For scholars, ignoring trade-offs can lead to underdeveloped theory, misspecified research designs, and flawed interpretations of empirical results. The trade-offs that are developed here need to be incorporated into theories of conflict onset, conflict escalation, and military effectiveness, among others. The biggest threat to research design posed by neglecting trade-offs is crafting flawed measures of military effectiveness itself. And, scholars risk misinterpreting observed empirical relationships if they are not prepared to look for the variety of possible relationships suggested by the trade-offs framework.

Incomplete accounts that neglect possible trade-offs also present dangers to policy makers. To be sure, policy makers sometimes comprehend and accept trade-offs, such as when elected leaders are forced to accept reductions in military effectiveness to maximize force protection, or when leaders take the calculated gamble of embracing an aggressive military strategy that risks conflict escalation. That said, for a variety of reasons policy makers sometimes fail to forecast trade-offs accurately, in turn causing them to draw false conclusions about the advisabilities of their policy choices. Insufficient experience with a particular mode of conflict or policy choice (such as a new weapons technology) can cause leaders to miss potential trade-offs. Belligerents sometimes fail to assess military performance effectively during wartime, in turn potentially undermining their ability to recognize the occurrence of trade-offs.[60]

[57] Berman et al. (2011a, 2011b). [58] Postel (2013).
[59] www.cbsnews.com/news/rumsfeld-skeptical-of-intel-analysis/.
[60] Blanken et al. (2015).

Psychological biases can come into play if, for example, the overvaluation of past success cause decision makers to downplay or miss entirely potential trade-offs.[61] The American failure in the Vietnam War was largely caused by overconfidence in the utility of massive firepower in defeating insurgents, and insufficient recognition that such firepower incurs a trade-off of alienating the local population. This error in turn was driven both by relative inexperience in fighting (recent) insurgent conflicts, as well as assuming that the World War II experience of winning through massive firepower could be repeated in the jungles of Southeast Asia.[62] Furthermore, even when policy makers try to assess trade-offs, they often do so poorly. Jeffrey Friedman in his chapter in this volume describes how policy makers' aversion to assigning quantitative probabilities to the likelihood of outcomes given certain strategies directly leads to poor decisions.

Exaggerated Trade-offs?

Though decision makers sometimes seek to improve military effectiveness and fail to grasp the potential trade-offs, the reverse also sometimes happens, of decision makers forgoing an opportunity to improve military effectiveness because of the exaggerated fear of some imagined trade-off. One example is the gender integration of the military. For years, decades, centuries, and even millennia, belligerents have largely excluded women from military service, especially in combat roles, mainly emerging from a view that women were not physically or psychologically fit for combat, and that their inclusion might undermine small unit cohesion.[63]

However, in recent years, decision makers in many countries have begun to take a broader view of the possible roles women might play in combat. They have observed that the average woman suffers only limited physiological inferiority in comparison to the average man (mostly upper body strength), and that many roles in modern militaries (including some combat roles, such as driving tanks and piloting aircraft) do not require high levels of performance in these physiological areas. Decision makers have begun to recognize that female soldiers/mixed-gender units may be even be *more* effective than male

[61] Kahneman et al. (1982). [62] Krepinevich (1986). [63] Goldstein (2001).

soldiers/single-gender units for at least some types of military operations, including peacekeeping and counterinsurgency.[64]

Militaries have also sometimes excluded or limited the combat opportunities of ethnic groups, such as American military exclusions of African Americans before the Korean War. Militaries fear that such groups are incapable of fighting well, the loyalty of such groups is suspect, and/or that their inclusion would undermine troop morale and cohesion.[65] The ability of many of these groups, such as Japanese American soldiers in World War II and African American soldiers throughout American history, to fight well demonstrates that such feared trade-offs do not always arise.

Informing Policy

It is important for scholars to develop the insight that the pursuit of military effectiveness can incur trade-offs, and to use that insight to inform policy.[66] This means that scholars should go beyond focusing on single causal relationships of a certain factor affecting one element of military effectiveness, and should also think about the big picture. Specifically, scholars will have specialized knowledge to build a comprehensive view, and have the ability to provide broader perspectives beyond analyzing single, causal relationships. This is especially true in the twenty-first century, as some scholars outside of government are becoming deeply informed about specific military effectiveness areas such as cyberwarfare, autonomous weapons systems, and gender integration of the military. These scholars sometimes have the ability not only to answer questions about cost and the diverse military effectiveness consequences of policy options, but they may even be uniquely suited to suggest implications that policy makers without their expertise might not have considered.

In the past, scholars and scientists have been on different occasions more or less willing to engage in broader debates about various elements of military effectiveness and foreign policy more broadly, sometimes affecting policy decisions, sometimes not. Some of the scientists involved in the Manhattan Project questioned the need for

[64] Karim and Beardsley (2017); Ricks (2009b); Denn (2014); Reiter (2015).

[65] Lyall (2016); Stam (1996).

[66] On how scholars can inform policy making, see Byman and Kroenig (2016).

using atomic bombs against Japan after the German surrender, encouraging the Truman administration to consider alternative courses of action such as first conducting a demonstration nuclear test.[67] Since 1945, nuclear scientists have continued to engage in broader policy debates, sometimes with positive policy consequences such as building the transnational scientific networks that helped advance nuclear arms control in the last years of the Cold War.[68]

Scholars have sometimes used trade-off insights to present big picture policy recommendations that proved to be accurate and insightful. As early as 1961, the political scientist Hans Morgenthau launched a comprehensive and unsparing critique of the American decision to intervene in Vietnam, correctly describing the conflict there as more anticolonial and nationalist than Communist, forecasting that even a unified, Communist Vietnam would not fall into the Chinese orbit, and pointing out that Vietnam had only marginal influence on the global material balance of power.[69] In the months before the 2003 Iraq War, international relations scholars made a unified set of claims that proved to be prescient, including that there were no ties between al-Qaeda and Iraq, the 2003 Iraq War would jeopardize the campaign against al-Qaeda, the US lacked an exit strategy, and the US would likely face years of occupation and policing after Saddam Hussein had been toppled.[70] Conversely, scholars need to watch carefully how decision makers selectively import ideas from academia, strategically using these ideas to justify certain foreign policy choices. For example, some scholars spoke out in opposition when the George W. Bush administration used the democratic peace proposition to justify the 2003 invasion of Iraq.[71]

This issue of scholarly engagement in the big policy picture is not limited to foreign policy. In the physical sciences, climatologists have had to decide how to engage the important public debate on global climate change. There is a temptation for some climate scientists to restrict their commentary to purely scientific and descriptive matters, such as reporting data on observed temperature changes. However, climate scientists have often gone beyond such comments, registering their opinions on key questions such as whether the observed temperature changes are likely human-made or natural, and what policy

[67] Frank (1999). [68] Evangelista (1999). [69] Zambernardi (2011).
[70] http://mearsheimer.uchicago.edu/pdfs/P0012.pdf. [71] Owen (2005).

solutions should be pursued. One motivation for their sustained and deep engagement is their belief that their reportage of scientific findings, such as climate observations and modeling results, are routinely misinterpreted by politicians, the public, and the media.[72]

To be clear, this is not to say that scholars should shy away from narrow, deep studies on single military effectiveness causal pathways. However, it is a call for scholars to recognize the dangers that narrow research can be used to support unwise policy. Scholars often underestimate the breadth of questions they have the ability to speak to, and the degree to which decision makers follow at least some forms of academic scholarship.[73] Most recently, the mid 2000s revamping of American counterinsurgency strategy, culminating in the highly successful 2007 Surge in Iraq and strategy shift toward population-centric counterinsurgency, was deeply informed by American military consultations with a wide array of scholars, especially social scientists.[74]

Road Map for the Book

The chapters in this volume strive to accomplish two objectives. First, they provide new ideas concerning both established and new areas of military effectiveness, providing theoretical, empirical, and policy insights. Second, and relatedly, they frame their discussions of military effectiveness around the trade-offs insight, demonstrating the ubiquity and significance of the trade-offs proposition. They also demonstrate similarities in trade-offs across different areas of military effectiveness.

Emanuele Castelli and Lorenzo Zambernardi in their chapter consider the imperatives of force protection, how some belligerents strive to minimize friendly military casualties during war. Force protection has become especially salient for casualty-sensitive modern democracies, as manifested by a greater emphasis on air strikes over ground troops, a preference for counterinsurgent forces to ride in armored vehicles rather than patrol on foot, and an increasing interest in drones and military robotics. Their chapter describes the evolution of force protection over time, with an emphasis on its role in war in recent

[72] Guardian (2014).
[73] In a survey, Avey and Desch (2014) found that American national security decision makers do follow academic scholarship, though with much greater interest in work using qualitative theoretical and empirical approaches.
[74] Ricks (2009a).

decades. Its central contribution is to develop the first set of arguments comprehensively assessing the connections between force protection and military effectiveness. Specifically, it describes three modes of force protection (and their accompanying trade-offs), *airpower*, *deployment without employment*, and *ground warfare force protection*. The chapter uses several historical episodes, especially from contemporary conflicts, to demonstrate these concepts.

In her chapter, Rosella Cappella Zielinski develops a framework for considering how war finance choices affect military effectiveness. The chapter builds a conceptual framework by proposing that belligerents have three sets of approaches to financing wars. *Direct* financing refers to means such as taxation, enforced savings, and enforced labor. *Indirect* financing refers to other domestic sources, especially debt. *External* financing refers to support received from outside of the state, such as third-party aid or plunder. Each approach presents different trade-offs in military effectiveness, and can be more or less feasible under a variety of different political or economic conditions. The theoretical framework developed in the chapter is applied to case studies of the US in the Korean War and Israel in the Yom Kippur War.

In his chapter, Jason Lyall describes how militaries use special "blocking" units to punish troops who defect and desert during battle, as a means of bolstering effectiveness. The chapter demonstrates the frequent use of this tactic, noting that since 1800 some 25 percent of belligerents in war have employed such units, and that this practice continues up to the present day, as belligerents such as ISIS in Iraq and Russian forces in Ukraine have deployed blocking units. The chapter develops the first conceptual framework explaining the complex effects such units have on military effectiveness. Belligerents deploy blocking units to improve effectiveness by maximizing the number of troops in battle by reducing the flow of deserters and defectors. However, such units can also create trade-offs. Manning the blocking units draws off troops from combat formations. Blocking units can sometimes be excessive in their punishment actions, executing or imprisoning large numbers of combat troops. Blocking units can also undermine troop morale. The chapter tests these ideas with case studies of the use of blocking units by the Soviet Army in World War II, at the battles of Stalingrad and Kursk.

Joseph Felter's chapter explores counterinsurgency, specifically the relationship between troop type and counterinsurgent effectiveness.

This chapter describes three categories of counterinsurgent troops: local, regular military, and special operations. It then describes why these different types of troops will experience different levels of counterinsurgent military effectiveness based on variance in qualities such as small unit leadership, training and doctrine, and decentralization of control. The theory proposes that special operations troops are particularly likely to be effective. The chapter then tests its theory using new, original data on insurgent violence in the Philippines over the time period 2001–2008. The chapter also describes possible trade-offs in reliance on special operations forces to conduct counterinsurgency operations. It develops a number of novel propositions, such as that increasing training to help troops become special operations forces may give those troops greater capacity to overthrow the government. Furthermore, training troops in counterinsurgency may make it more difficult to train them to fight in conventional warfare.

Michael Horowitz in his chapter considers the military effectiveness implications of a relatively new element of violent conflict, military robotic systems such as drones and self-guided missiles. The chapter first describes the array of military robotics systems, some deployed, some in development, and others still on the drawing board. The chapter then develops a conceptual structure for thinking about the military effectiveness issues associated with military robotics. For example, one of the clearest advantages of military robotics is force protection, the ability to engage in combat without risking friendly casualties. The chapter also works through other important, but less recognized military effectiveness implications of military robotics, such as command and control, vulnerability to cyberattacks, and flight ranges and endurance of drones, among others.

Caitlin Talmadge's chapter considers the linkage between conventional military effectiveness and the shadow of nuclear escalation. Though nuclear weapons play a critical role in international relations, they have been surprisingly absent in discussions of military effectiveness. This chapter provides some of the first analysis of how possession of nuclear weapons affects conventional military effectiveness. It argues that when nuclear-armed states fight conventional wars, there is the possibility of escalation to nuclear use, either by choice or inadvertently. Highly successful conventional military operations against a nuclear-armed state create risks of nuclear escalation, meaning that in the presence of nuclear weapons high levels of conventional military effectiveness may present very dangerous trade-offs. The chapter then applies its conceptual argument to a number of historical cases, including the India–Pakistan conflict, the Yom Kippur War, and the Korean War. The chapter also applies the argument to areas of policy interest, such as the possibility of inadvertent nuclear escalation in the event of China–US naval clashes over disputes in the South China Sea or the security of Taiwan, or expanded Russian aggression in Europe.

In his chapter, Jeffrey Friedman develops a different kind of argument, showing how military officers often fail to present the appropriate kinds of probability assessments to their civilian leaders, with important implications for crafting optimal military and foreign policy choices. Put differently, while other chapters examine the *existence* of trade-offs, this chapter examines how decision makers *assess* trade-offs. Building on decision theory, the chapter argues that military decision makers routinely present civilian decision makers with vague, imprecise, and sometimes misleading probability estimates of different outcomes, making foreign policy fiascoes more likely. The core argument is then demonstrated in a number of US foreign policy decisions, such as the 1961 Bay of Pigs invasion, the escalation of the Vietnam War, and the 2009 Surge in the Afghanistan War. The chapter also recommends best decision-making practices and how military leaders should present their beliefs about the probabilities that different outcomes would flow from different policy choices.

Filippo Andreatta's chapter concludes, offering critical insights into the trade-offs argument as well as suggestions for future research.

References

Andreski, Stanislav. 1980. On the Peaceful Disposition of Military Dictatorships. *Journal of Strategic Studies* 3 (December): 3–10.

Arreguín-Toft, Ivan. 2005. *How the Weak Win Wars: A Theory of Asymmetric Conflict*. Cambridge: Cambridge University Press.

Atkinson, Rick. 2007. *The Day of Battle: The War in Sicily and Italy, 1943–1944*. New York: Picador.

Avant, Deborah. 1994. *Political Institutions and Military Change: Lessons from Peripheral Wars*. Ithaca, NY: Cornell University Press.

Avey, Paul C., and Michael C. Desch. 2014. What Do Policymakers Want from Us? Results of a Survey of Current and Former Senior National Security Decision Makers. *International Studies Quarterly* 58 (June): 227–246.

Bartov, Omer. 1992. *Hitler's Army: Soldiers, Nazis, and War in the Third Reich*. Oxford: Oxford University Press.

Beckley, Michael C. 2010. Economic Development and Military Effectiveness. *Journal of Strategic Studies* 33(February): 43–79.

Berinsky, Adam J. 2009. *In Time of War: Understanding American Public Opinion from World War II to Iraq*. Chicago: University of Chicago Press.

Berman, Eli, Michael Callen, Joseph H. Felter, and Jacob N. Shapiro. 2011a. Do Working Men Rebel? Insurgency and Unemployment in Afghanistan, Iraq, and the Philippines. *Journal of Conflict Resolution* 59(4): 496–528.

Berman, Eli, Jacob N. Shapiro, and Joseph H. Felter. 2011b. Can Hearts and Minds Be Bought? The Economics of Counterinsurgency in Iraq. *Journal of Political Economy* 119(August): 766–819.

Biddle, Stephen. 2004. *Military Power: Explaining Victory and Defeat in Modern Battle*. Princeton, NJ: Princeton University Press.

Biddle, Stephen, and Stephen Long. 2004. Battlefield Military Effectiveness: A Deeper Look. *Journal of Conflict Resolution* 48(4): 525–546.

Biddle, Stephen, and Robert Zirkle. 1996. Technology, Civil-Military Relations, and Warfare in the Developing World. *Journal of Strategic Studies* 19(June): 171–212.

Blanken, Leo J., Hy Rothstein, and Jason J. Lepore. 2015. *Assessing War: The Challenge of Measuring Success and Failure*. Washington, DC: Georgetown University Press.

Brodie, Bernard. 1973. *War and Politics*. New York: Macmillan.

Brooks, Risa A. 2008. *Shaping Strategy: The Civil-Military Politics of Strategic Assessment*. Princeton, NJ: Princeton University Press.

Brooks, Risa A., and Elizabeth A. Stanley. 2007. *Creating Military Power: The Sources of Military Effectiveness*. Stanford, CA: Stanford University Press.

Brown, Michael E., Owen R. Coté Jr., Sean M. Lynn-Jones, and Steven E. Miller, eds. 2011. *Do Democracies Win Their Wars?* Cambridge: Cambridge University Press.

Buckingham, William A., Jr. 1983. Operation Ranch Hand: Herbicides in Southeast Asia. *Air University Review* 34(5): 42–53.

Bueno de Mesquita, Bruce. 1981. *The War Trap*. New Haven, CT: Yale University Press.

Bueno de Mesquita, Bruce, Alastair Smith, Randolph M. Siverson, and James D. Morrow. 2003. *The Logic of Political Survival*. Cambridge, MA: MIT Press.

Byman, Daniel, and Matthew Kroenig. 2016. Reaching beyond the Ivory Tower: A How to Manual. *Security Studies* 25(2): 289–319.

Cappella Zielinski, Rosella. 2016. *How States Pay for Wars*. Ithaca, NY: Cornell University Press.

Castillo, Jasen. 2014. *Endurance and War: The National Sources of Military Cohesion*. Stanford, CA: Stanford University Press.

Caverley, Jonathan D. 2014. *Democratic Militarism: Voting, Wealth, and War*. Cambridge: Cambridge University Press.

Chiozza, Giacomo, and H. E. Goemans. 2011. *Leaders and International Conflict*. Cambridge: Cambridge University Press.

Choi, Ajin. 2003. The Power of Democratic Cooperation. *International Security* 28(Summer): 142–153.

2012. Fighting to the Finish: Democracy and Commitment in Coalition War. *Security Studies* 21(4): 624–653.

Clausewitz, Carl, von. 1976. *On War*. Michael Howard and Peter Paret, trans. and ed. Princeton, NJ: Princeton University Press.

Clodfelter, Michael. 2008. *Warfare and Armed Conflicts: A Statistical Encyclopedia of Casualty and Other Figures, 1494–2007*. Jefferson, NC: McFarland.

Croco, Sarah E. 2015. *Peace at What Price? Leaders and the Domestic Politics of War Termination*. Cambridge: Cambridge University Press.

Denn, William. 2014. Women in Combat Roles Would Strengthen the Military. *Washington Post*, April 3.

Desch, Michael C. 2008. *Power and Military Effectiveness: The Fallacy of Democratic Triumphalism*. Baltimore: Johns Hopkins University Press.

Dower, John W. 1986. *War without Mercy: Race and Power in the Pacific War*. New York: Pantheon.

Downes, Alexander B. 2008. *Targeting Civilians in War*. Ithaca, NY: Cornell University Press.

Evangelista, Matthew. 1988. *Innovation and the Arms Race: How the United States and the Soviet Union Develop New Military Technologies*. Ithaca, NY: Cornell University Press.

1999. *Unarmed Forces: The Transnational Movement to End the Cold War*. Ithaca, NY: Cornell University Press.

Flores-Macías, Gustavo A., and Sarah E. Kreps. 2013. Political Parties at War: A Study of American War Finance, 1789–2010. *American Political Science Review* 107(November): 833–848.

Forthcoming. Borrowing Support for War: The Effect of War Finance on Public Attitudes toward Conflict. *Journal of Conflict Resolution*.

Frank, Richard B. 1999. *Downfall: The End of the Imperial Japanese Empire*. New York: Random House.

Gartner, Scott Sigmund. 1997. *Strategic Assessment in War*. New Haven, CT: Yale University Press.

2008. The Multiple Effects of Casualties on Public Support for War: An Experimental Approach. *American Political Science Review* 102(February): 95–106.

Gartner, Scott Sigmund, and Gary Segura. 1998. War, Casualties, and Public Opinion. *Journal of Conflict Resolution* 42(June): 278–300.

Gartner, Scott Sigmund, and Gary M. Segura. 2008. All Politics Are Still Local: The Iraq War and the 2006 Midterm Elections. *PS: Political Science & Politics* 41(1): 95–100.

Gartner, Scott Sigmund, Gary Segura, and Michael Wilkening. 1997. All Politics Are Local: Local Losses and Individual Attitudes toward the Vietnam War. *Journal of Conflict Resolution* 41(October): 669–694.

Gelpi, Christopher, Peter D. Feaver, and Jason Reifler. 2009. *Paying the Human Costs of War: American Public Opinion and Casualties in Military Conflicts*. Princeton, NJ: Princeton University Press.

Gilbert, Felix. 1986. Machiavelli: The Renaissance of the Art of War. In Peter Paret, ed., *Makers of Modern Strategy from Machiavelli to the Nuclear Age*, 11–31. Princeton, NJ: Princeton University Press.

Goemans, H. E. 2000. *War and Punishment: The Causes of War Termination and the First World War*. Princeton, NJ: Princeton University Press.

Goldman, Emily O., and Leslie C. Eliason, eds. 2003. *The Diffusion of Military Technology and Ideas*. Stanford, CA: Stanford University Press.

Goldstein, Joshua. 2001. *War and Gender*. Cambridge: Cambridge University Press.

Grauer, Ryan. 2014. Why Do Soldiers Give Up? A Self-Preservation Theory of Surrender. *Security Studies* 23(3): 622–655.

Grauer, Ryan, and Michael C. Horowitz. 2012. What Determines Military Victory? Testing the Modern System. *Journal of Conflict Resolution* 21(1): 83–112.

Guardian. 2014. What Really Annoys Scientists about the State of the Climate Change Debate? June 25. www.theguardian.com/environment/planet-oz/2014/jun/26/what-really-annoys-scientists-about-the-state-of-the-climate-change-debate.

Hallek, Robert. 1998. *Flawed Giant: Lyndon and His Times, 1961–1973*. New York: Oxford University Press.

Hasegawa, Tsuyoshi. 2005. *Racing the Enemy: Stalin, Truman, and the Surrender of Japan*. Cambridge, MA: Belknap Press.

Herring, George C. 1996. *America's Longest War: The United States and Vietnam, 1950–1975*. 3rd ed. New York: McGraw-Hill.

Horowitz, Michael C. 2010. *The Diffusion of Military Power: Causes and Consequences for International Politics*. Princeton, NJ: Princeton University Press.

Horowitz, Michael C., Erin Simpson, and Allan C. Stam. 2011. Domestic Institutions and Wartime Casualties. *International Studies Quarterly* 55(4): 909–936.

Horowitz, Michael C., and Dan Reiter. 2001. When Does Aerial Bombing Work? Quantitative Empirical Tests, 1917–1999. *Journal of Conflict Resolution* 45(April): 147–173.

James, Patrick, and Seung-Whan Choi. 2005. *Civil-Military Dynamics, Democracy, and International Conflict: A New Quest for International Peace*. New York: Palgrave.

Jervis, Robert. 1978. Cooperation under the Security Dilemma. *World Politics* 30(January): 167–214.

Johnston, Patrick B. 2012. Does Decapitation Work? Assessing the Effectiveness of Leadership Targeting in Counterinsurgency Campaigns. *International Security* 36(Spring): 47–79.

Johnston, Patrick B., and A. Sarbahi. Forthcoming. The Impact of US Drone Strikes on Terrorism in Pakistan and Afghanistan. *International Studies Quarterly*.

Jordan, Jenna. 2009. When Heads Roll: Assessing the Effectiveness of Leadership Decapitation. *Security Studies* 18(4): 719–755.

Kahneman, Daniel, Paul Slovic, and Amos Tversky. 1982. *Judgment under Uncertainty: Heuristics and Biases*. Cambridge: Cambridge University Press.

Kalyvas, Stathis N. 2006. *The Logic of Violence in Civil War*. Cambridge: Cambridge University Press.

Kaplan, Lawrence S. 2006. McNamara, Vietnam, and the Defense of Europe. In Vojtech Mastny, Sven S. Holtsmark, and Andreas Wenger, eds., *War Plans and Alliances in the Cold War: Threat Perceptions in the East and West*, 286–300. New York: Routledge.

Karim, Sabrina, and Kyle Beardsley. 2017. *Equal Opportunity Peacekeeping: Women, Peace, and Security in Post-Conflict States*. Oxford: Oxford University Press.

Khong, Yuen Foong. 1992. *Analogies at War: Korea, Munich, Dien Bien Phu, and the Vietnam Decisions of 1965*. Princeton, NJ: Princeton University Press.

Kier, Elizabeth. 1997. *Imagining War: French and British Military Doctrine between the Wars*. Princeton, NJ: Princeton University Press.

King, Anthony. 2016. On Combat Effectiveness in the Infantry Platoon: Beyond the Primary Group Thesis. *Security Studies* 25(4): 699–728.

Kocher, Matthew Adam, Thomas B. Pepinsky, and Stathis N. Kalyvas. 2012. Aerial Bombing and Counterinsurgency in the Vietnam War. *American Journal of Political Science* 55(April): 201–218.

Krepinevich, Andrew F., Jr. 1986. *The Army and Vietnam*. Baltimore: Johns Hopkins University Press.

Kriner, Douglas L., Breanna Lechase, and Rosella Cappella Zielinski. Forthcoming. Self-Interest, Partisanship, and the Conditional Influence of Taxation on Support for War in the United States. *Conflict Management and Peace Science*.

Lake, David A. 1992. Powerful Pacifists: Democratic States and War. *American Political Science Review* 86(March): 24–37.

Levi, Margaret. 1997. *Consent, Dissent, and Patriotism*. Cambridge: Cambridge University Press.

Long, Austin. 2014. Whack-a-Mole or Coup de Grace? Institutionalization and Leadership Targeting in Iraq and Afghanistan. *Security Studies* 23(3): 471–512.

Lyall, Jason. 2009. Does Indiscriminate Violence Incite Insurgent Attacks? Evidence from Chechnya. *Journal of Conflict Resolution* 53(June): 331–362.

2010. Are Coethnics More Effective Counterinsurgents? Evidence from the Second Chechen War. *American Political Science Review* 104(February): 1–20.

2016. Paths to Ruin: Explaining Military Effectiveness in Conventional War. Unpublished manuscript, Yale University.

Lyall, Jason, and Isaiah Wilson III. 2009. Rage against the Machines: Explaining Outcomes in Counterinsurgency Wars. *International Organization* 63(Winter): 67–106.

McPherson, James M. 1988. *Battle Cry of Freedom: The Civil War Era*. New York: Ballantine Books.

Mearsheimer, John J. 1983. *Conventional Deterrence*. Ithaca, NY: Cornell University Press.

Merom, Gil. 2003. *How Democracies Lose Small Wars: State, Society and the Failures of France in Algeria, Israel in Lebanon, and the United States in Vietnam*. Cambridge: Cambridge University Press.

Millett, Allan R., Williamson Murray, and Kenneth H. Watman. 1988. The Effectiveness of Military Organizations. In Allan R. Millett and Williamson Murray, eds., *Military Effectiveness*, vol. 1, 1–30. Boston: Allen and Unwin.

Mueller, John. 1973. *War, Presidents and Public Opinion*. New York: John Wiley.

Organski, A. F. K., and Jacek Kugler. 1980. *The War Ledger*. Chicago: University of Chicago Press.

Overy, Richard. 1995. *Why the Allies Won*. New York: W. W. Norton.

Owen, John M., IV. 2005. Iraq and the Democratic Peace. *Foreign Affairs*, November/December.

Pape, Robert A. 1996. *Bombing to Win*. Ithaca, NY: Cornell University Press.

Peic, Goran. 2014. Civilian Defense Forces, State Capacity, and Government Victory in Counterinsurgency Wars. *Studies in Conflict and Terrorism* 37(2): 162–184.

Peled, Alon. 1998. *A Question of Loyalty: Military Manpower Policy in Multiethnic States*. Ithaca, NY: Cornell University Press.

Pilster, Ulrich, and Tobias Böhmelt. 2011. Coup-Proofing and Military Effectiveness in Interstate Wars. *Conflict Management and Peace Science* 28(4): 331–350.

2012. Do Democracies Engage in Less Coup-Proofing? On the Relationship between Regime Type and Civil-Military Relations. *Foreign Policy Analysis* 8(October): 355–372.

Poast, Paul. 2015. Central Banks at War. *International Organization* 69(Winter): 63–95.

Postel, Thérèse. 2013. How Guantanamo Bay's Existence Helps al-Qaeda Recruit More Terrorists. *Atlantic*, April 12.

Price, Bryan C. 2012. Targeting Top Terrorists: How Leadership Decapitation Contributes to Counterterrorism. *International Security* 36 (Spring): 9–46.

Price, Richard M. 1997. *The Chemical Weapons Taboo*. Ithaca, NY: Cornell University Press.

Quinlivan, James T. 1999. Coup-Proofing: Its Practice and Consequences in the Middle East. *International Security* 24(Fall): 131–165.

Reiter, Dan. 1995. Exploding the Powder Keg Myth: Preemptive Wars Almost Never Happen. *International Security* 20(Fall): 5–34.

2007. Nationalism and Military Effectiveness. In Elizabeth A. Stanley-Mitchell and Risa Brooks, eds., *Creating Military Power: The Sources of Military Effectiveness*, 27–54. Palo Alto, CA: Stanford University Press.

2009. *How Wars End*. Princeton, NJ: Princeton University Press.

2015. The Positivist Study of Gender and International Relations. *Journal of Conflict Resolution* 59(7): 1301–1326.

Reiter, Dan, and Curtis Meek. 1999. Determinants of Military Strategy, 1903–1994: A Quantitative Empirical Test. *International Studies Quarterly* 43(June): 363–387.

Reiter, Dan, and Allan C. Stam. 2002. *Democracies at War*. Princeton, NJ: Princeton University Press.

Reiter, Dan, and William Wagstaff. Forthcoming. Leadership and Military Effectiveness. *Foreign Policy Analysis*.

Resende-Santos, Joao. 2007. *Neorealism, States, and the Modern Mass Army*. Cambridge: Cambridge University Press.

Ricks, Thomas E. 2006. *Fiasco: The American Military Adventure in Iraq*. New York: Penguin.

2009a. *The Gamble: General Petraeus and the American Military Adventure in Iraq*. New York: Penguin.

2009b. Women in COIN (II): How to Do It Right. *Foreign Policy*, October 9. http://foreignpolicy.com/2009/10/09/women-in-coin-ii-how-to-do-it-right/.

Risse-Kappen, Thomas. 1995. *Cooperation among Democracies: The European Influence on US Foreign Policy*. Princeton, NJ: Princeton University Press.

Rosen, Stephen Peter. 1991. *Winning the Next War: Innovation and the Modern Military*. Ithaca, NY: Cornell University Press.

1995. Military Effectiveness: Why Society Matters. *International Security* 19(Spring): 5–31.

1996. *Societies and Military Power: India and Its Armies*. Ithaca, NY: Cornell University Press.

Rotte, Ralph, and Christoph Schmidt. 2003. On the Production of Victory: Empirical Determinants of Battlefield Success in Modern War. *Defence and Peace Economics* 14(3): 175–192.

Russett, Bruce M. 1993. *Grasping the Democratic Peace: Principles for a Post–Cold War World*. Princeton, NJ: Princeton University Press.

Schultz, Kenneth A., and Barry Weingast. 2003. The Democratic Advantage: Institutional Foundations of Financial Power in International Competition. *International Organization* 57(1): 3–42.

Shea, Patrick E. Forthcoming. Financing Victory: Sovereign Credit, Democracy, and War. *Journal of Conflict Resolution*.

Shils, Edward A., and Morris Janowitz. 1948. Cohesion and Disintegration in the Wehrmacht in World War II. *Public Opinion Quarterly* 12(2): 280–315.

Snyder, Jack. 1984. *Ideology of the Offensive: Military Decision Making and the Disasters of 1914*. Ithaca, NY: Cornell University Press.

Stam, Allan C., III. 1996. *Win, Lose, or Draw: Domestic Politics and the Crucible of War*. Ann Arbor: University of Michigan Press.

Sullivan, Patricia. 2012. *Who Wins? Predicting Strategic Success and Failure in Armed Conflict*. Oxford: Oxford University Press.

Talmadge, Caitlin. 2015. *The Dictator's Army: Battlefield Effectiveness in Authoritarian Regimes*. Ithaca, NY: Cornell University Press.

Tannenwald, Nina. 1999. The Nuclear Taboo: The United States and the Normative Basis of Nuclear Non-Use. *International Organization* 53(Summer): 433–468.

Tuchman, Barbara. 1962. *The Guns of August*. New York: Bantam.

Vasquez, Joseph Paul. 2005. Shouldering the Soldiering: Democracy, Conscription, and Military Casualties. *Journal of Conflict Resolution* 49(6): 849–873.

Weeks, Jessica L. P. 2014. *Dictators at War and Peace*. Ithaca, NY: Cornell University Press.

Zambernardi, Lorenzo. 2011. The Impotence of Power: Morgenthau's Critique of American Intervention in Vietnam. *Review of International Studies* 37(July): 1335–1356.

2 Force Protection and Its Trade-offs

EMANUELE CASTELLI AND
LORENZO ZAMBERNARDI

"Soldiers are made to be killed." With these words, not only did Napoleon voice his personal, cynical view of soldiers[1] but he also gave a rather accurate description of how commoners were perceived by statesmen and generals for the most part of modern and contemporary history. Soldiers had been dying in wars for centuries, and with the exception of humanist thinkers such as Erasmus, Saint-Pierre, Kant, Tolstoy, and a few other distinguished figures, no one had really complained about that. Apart from the purely strategic point of view, there is no evidence that the problem of soldiers' expendability had any influence on the conduct of war. Soldiers were of value as "fighting commodities,"[2] not as human beings. The primary motive to salvage the lives of soldiers in battle lay in the utilitarian fact that more soldiers implied greater combat power.

Over the last decades, however, protection of troops seems to have become increasingly important for Western states. The traditional attitude in which soldiers' deaths were politically and morally unproblematic provides a striking contrast to the present understanding, where protecting the lives of combatants seems to be a necessary requirement for their employment on the battlefield. Indeed, it appears evident that when countries like the US, Britain, France, Germany, Italy, and many others employ coercive force abroad, they do their best to reduce lethal risks to their combatants. Thus, while for centuries states employed technology to make mass slaughter possible, today they use it also to minimize friendly casualties. In military jargon this policy is generally termed *force protection*. Although this term is usually employed to describe a set of tactical micromeasures intended to reduce soldiers' exposure to enemy fire (i.e., checkpoint protocols, decisions on the

[1] In Walzer (2006 [1977], 136); Summerville (2005, 4).
[2] Showalter and Astore (2007, 133). To be sure, dying for one's country on the battlefield could be even considered as a glorious, honorable sacrifice for the nation.

threshold for using coercive force, etc.), here we use it in a broader way to refer also to all those macromilitary policies which are employed to create a safe distance between Western soldiers and enemy combatants such as the use of drones, long-distance artillery, mechanized warfare, and the like.

So far, scholars of war and international relations have treated force protection either as a dependent variable resulting from casualty sensitivity[3] or as an independent variable accounting for the incapacity of discriminating between civilians and enemy combatants in the warzone.[4] In doing so, scholars have succeeded in covering both some of the consequences of force protection and some of its main ethical implications, but they have so far disregarded an interesting line of inquiry, i.e., the study of its influence on military effectiveness.[5] By considering a number of possible trade-offs originating from the adoption of force protection measures, this essay is intended to fill this gap. In particular, we argue that force protection trades off with military effectiveness in irregular conflict. Indeed, as a large body of scholarship has shown, increasing friendly military casualties tend to erode public support for war, forcing a belligerent either to exit the war short of victory or to employ military means intended to minimize military dead. In turn, as we will show in this paper, many attempts to reduce casualties produce political, security, and war-fighting trade-offs, which result in a lower level of military effectiveness.

First, there are political trade-offs: protecting military forces may push states to adopt military policies (such as airpower or heavy mechanization strategies) that increase the physical distance between their own troops and enemy forces. In turn, this may make it difficult to discriminate between combatants and noncombatants, thus increasing enemy civilian casualties and undermining local support for counterinsurgency forces. This is especially true in irregular warfare, as every policy that attempts to minimize friendly casualties often escalates local civilian casualties, jeopardizing the fundamental goal of counterinsurgency strategy, i.e., securing the populace. Second, the trade-off between force protection and military effectiveness may produce new security threats. On the one hand, besides alienating local support,

[3] Cornish (2003); Levy (2009a; 2010; 2012); Zambernardi (2010).
[4] Shaw (2002; 2005); T. W. Smith (2008); Carpenter (2011).
[5] For analyses of military effectiveness, see, e.g., Millet et al. (1988); Reiter (2007).

indiscriminate bombing may fuel the insurgency, bringing new recruits to rebel organizations; on the other, subcontracting fighting to local militias – that is, training and providing them with weapons and equipment – may increase the risk that they may decide to revolt against their supporters. Third and lastly, there are war-fighting trade-offs: heavy reliance on indigenous forces, for example, entails problems of troop control and discipline, affecting the extent and the quality of their performance. Moreover, the use of airpower and mechanized means in irregular conflicts reduces military effectiveness by limiting the capacity of collecting reliable information and intelligence on insurgent groups.

This chapter is organized in two main sections. In the first part we briefly clarify what force protection is and how it has changed over time. This study shows that force protection can be driven by two different goals: first, it may simply be instrumental to the need of preserving combat power (i.e., having more soldiers generally means more military might); second, it may be driven by the goal of preserving soldiers' lives as an end in itself, regardless of strategic and tactical considerations. This is a crucial distinction, since this second driver of force protection, unlike the first,[6] is not only a relatively new phenomenon, but, most importantly, it generates a variety of trade-offs impacting on military effectiveness. In the second part, after having described some military practices designed for force protection (i.e., a heavy reliance on airpower and indigenous forces, and a series of tactical measures that reduce soldiers' exposure during ground operations), we discuss some of the most important trade-offs resulting from these practices, with special reference to irregular warfare. Finally, the conclusion outlines why force protection in the foreseeable future is likely to remain a source of significant trade-offs in terms of military effectiveness.

The Changing Nature of Force Protection

Throughout history, soldiers' deaths were mainly seen as an inevitable, intrinsic aspect of war. This is especially true for Western armies, since the preference for highly lethal pitched battles aimed at destroying opponent's forces was one of the primary features of the so-called Western way of warfare.[7] Such an attitude was cynically embodied

[6] See, e.g., Mearsheimer (1983); Biddle (2004). [7] Hanson (1989).

by Napoleon, who famously declared that "he could replace 30,000 men a month."[8] Of course, this does not equate to suggesting that protection of forces was unimportant in battle. On the contrary, preserving armed forces was a "normal military behavior,"[9] in the sense that it was instrumental to the need to preserve fighting power for striking enemy forces. As Clausewitz noted, "destroying the enemy's forces [requires] the preservation of your own" and "these two efforts always go together, they interact."[10] In other words, force protection was a mere prerequisite of military effectiveness: the more soldiers one had, the more likely it was that one could succeed in the bloody business of war.

With the gradual development of accurate, lethal firearms – thanks to improvements in weaponry and targeting – force protection became increasingly important and turned into a prerequisite for and an essential dimension of the effective use of coercive power. As Stephen Biddle put it, the increasing lethality of firepower "defined the central problem of modern tactics: how to survive the hail of metal long enough to perform meaningful military missions."[11]

In the course of the Great War, belligerents realized that not only had direct infantry offensives failed to achieve any concrete result but they had also produced an enormous drain of manpower and even mutinies, such as in the French troops. Accordingly, force dispersal on the battlefield became a crucial element and a "fundamental property of modern warfare,"[12] while reducing force exposure to enemy's lethal firepower became a key feature for successful military operations.[13] Twenty years later, by exploiting the advantages of mechanized warfare, Nazi Germany succeeded in achieving tactical effectiveness at low human costs. Indeed, the German maneuver-based strategy was not only militarily effective but it also produced few friendly casualties. As John Keegan and Richard Holmes[14] reminded us in comparing World War I with

[8] Rothenberg (2000, 112). [9] Smith (2005, 490).
[10] Clausewitz (1993, 112). [11] Biddle (2004, 30). [12] Biddle (2004, 31).
[13] However, the relationship between Biddle's concept of "modern system" and military effectiveness seems more nuanced than that. For example, according to Grauer and Horowitz (2012), who performed a test of Biddle's argument, the modern system of force employment (i.e., grounded on cover and concealment, dispersion, and the break through enemy lines) has not had an independent effect on battlefield outcomes. Other unit-level factors, such as regime type, social structure, culture, and the type of civil–military relations, should be taken into account to explain tactical victory.
[14] Keegan and Holmes (1985, 35).

the initial phase of World War II: "German casualties in a campaign that overwhelmed Belgium, Holland, and France were less than half of those the Kaiser's army had suffered at the battle of Verdun in 1916. Verdun had ended in stalemate; the 1940 campaign had made Germany master of the West."

The change in the lethality of firepower was followed by an independent – even though related – development. Indeed, since the end of World War I, military casualties have slowly but gradually become not only a military problem but also a highly sensitive political issue. This is what political scientists have termed casualty sensitivity. As a matter of fact, today the priority for Western armies is not only to be militarily effective – that is, accomplishing the goals assigned to them by political leaders at acceptable costs, but also to do it with few friendly casualties – to be efficient in terms of the human cost of war.[15] In turn, the necessity of minimizing casualties significantly shapes the ways force and troops are employed on the battlefield, constraining military options and making force protection a political requirement for Western militaries.

In the next section, we show that force protection trades off with some aspects of military effectiveness, especially in those irregular wars where discriminating between combatants and noncombatants is hardly possible. Indeed, these complex, multifaceted conflicts do not correspond to the traditional clash between regular armies.[16] Irregular warfare is usually characterized by the presence of small, mobile groups of combatants who do not carry arms openly and who conduct

[15] Although the origins of casualty sensitivity are not yet clear, and here we are not in the position to discuss the origins of this phenomenon, scholars have offered a variety of explanations, such as the gradual but irreversible decline of war enthusiasm (Mueller 1989; Pinker 2011), the recruitment system (i.e., from conscription to volunteer armies) (see Levy 2009b; Horowitz et al. 2011), and regime type (i.e., democracies tend to suffer fewer military casualties than their autocratic counterparts) (see Valentino et al. 2010). Although the simple association between mounting casualties and decline in domestic support for war has been challenged by a variety of scholars, who have contended that public attitudes toward foreign intervention result from a number of factors, such as the rhetorical strategy and the framing of war by the political leadership (Berinsky 2007, 2009), civil–military relations (Gelpi and Feaver 2004), and the perceived likelihood of success in war (Gelpi 2006; Gelpi et al. 2005, 2009), no one has really challenged the fact that sensitivity to casualties is greatly increased in the recent decades.

[16] On irregular, asymmetric, and unconventional warfare, see, e.g., Van Creveld (1991), Gray (2003), Kaldor (1998), Arreguín-Toft (2005), Münkler (2005), Hoffman (2007), R. Smith (2008), and Andreatta (2012).

unconventional military actions, such as terrorist attacks, ambushes, raids, and guerrilla operations. Irregular combatants avoid open confrontations with enemy forces and prefer to rely on "hit-and-run" tactics. Then, after carrying out these military actions, they either disguise as civilians or return to their safe havens. In the second place, since irregular combatants do not wear uniforms, one of the main differences between conventional and irregular warfare is that, in the former, the enemy forces can be detected and attacked, while in the latter, the enemy is much less recognizable. Thus, in irregular conflicts, civilians and insurgents are generally (though not always) intermingled and difficult to distinguish.

These typical features of irregular warfare make it difficult for the intervening state to strike the insurgents while protecting the population, which can be considered the Clausewitzian center of gravity of irregular conflict.[17] Indeed, the populace is the main source of loyalty and support (for the government), intelligence (for finding and fighting irregular combatants), and recruitment (in local anti-insurgency forces). Thus, acquiring the consensus of the local population is a crucial strategic goal.[18]

Seen more broadly, irregular warfare can be understood as a battle between two different centers of gravity. On the one hand, there is the local population, whose support has to be gained; on the other, there is the public of the intervening state, whose support has to be maintained. Both belligerents will try to hit the opponent's center of gravity: by inflicting military casualties, insurgents try to induce the occupying

[17] In his magisterial work, Clausewitz referred to the idea of "center of gravity" to illustrate how the enemy could be defeated. Drawing upon the analogy with physics, where the "center of gravity" is "always found where the mass is concentrated most densely" (Clausewitz 1993, 587), the Prussian general argued that "the same holds true in war" (587). Accordingly, to knock the enemy down, an army has to throw the enemy forces off balance. The possible centers of gravity in conventional warfare might include the enemy army, the capital city, or, when small powers rely on major ones, their political alliance (720). However, according to Clausewitz's logic, the center of gravity may change depending on the nature of the conflict. As we suggest below, in irregular warfare, the "hub of all power and movement, on which everything depends, that is the point against which all our energies should be directed" (720), is the local population.

[18] Even a harsh critic of American counterinsurgency, such as Gian Gentile (2013), does not dispute that securing the population is a crucial strategic requirement for fighting insurgencies. For critiques of counterinsurgency doctrine, see Griffin (2010); Strachan (2010).

forces to withdraw short of victory; by protecting and securing the populace, the occupying forces try to maintain the support of the locals. Losing that support, for both sides, may decide the war outcome.

Disaggregating Force Protection

In order to illustrate the trade-offs resulting from force protection and what they exactly imply in empirical terms, we need to disaggregate the concept of force protection and group its main manifestations into three main categories.[19] The first one is the preference for *airpower* in the broad sense of the term, therefore including not only air strikes but the use of drones and cruise missiles as well. The second is what can be termed *deployment without employment*, meaning that military troops may be deployed in the theater of war, but they are not used to engage in close combat with enemy forces (i.e., they can either stay in their highly protected compounds or subcontract fighting to indigenous forces). The third category concerns those practices that are used when friendly armed forces are employed in ground warfare. In this case, they adopt both mechanized warfare and a series of tactical micromeasures meant to reduce exposure to enemy fire. This is what we term *ground warfare force protection*. In the next three paragraphs, we discuss how this set of military practices may incur in trade-offs that significantly impact on military effectiveness.

Airpower as Force Protection

One way to minimize friendly casualties is by heavily relying on airpower, that is to say using air strikes, drones, and cruise missiles. As a matter of fact, airpower is employed not only to destroy enemy forces, but also to gain a safe distance between one's soldiers and enemy combatants. While strategic bombing against civilians and infrastructure during World War II was mainly intended both to break the enemy's will and its capacity to fight, since the Korean War the US has made an

[19] We are aware that the distinction between those categories may look artificial. In fact, armed forces use a combination of these three military practices and try to integrate multiple instruments and policies to increase effectiveness. However, by analyzing each of them individually, we can better illustrate their consequences in terms of trade-offs. The discussion of these three military practices is adapted from Zambernardi (2017).

extensive use of airpower with the deliberate goal of reducing friendly casualties.[20] Such a heavy reliance on airpower for minimizing the human cost of war is also evident during conflicts in the post–Cold War period. Indeed, it is no coincidence that in every recent major conflict, which has involved Western states, from Kosovo to Libya, the use of long-distance missiles launched by land, sea, and air has played the central role in military operations. More specifically, during operations, such as Deliberate Force (Bosnia and Herzegovina), Infinite Reach (Afghanistan and Sudan), Desert Fox (Iraq), Allied Force (Kosovo), and Unified Protector (Libya), the exclusive reliance on aerial firepower ensured that Western combatants were kept out of the war zone.

While the US provides the vast majority of capabilities employed in these bombing campaigns, European states appear to have become perfectly comfortable with this military approach. To be sure, at the time of the war in Kosovo, NATO's European members were opened to a ground operation, in contrast to the US, but in the following years they have moved closer toward a preference for aerial bombing, as clearly exemplified by the way they fought in Libya in order to stop Gaddafi's operations against civilians. Certainly, a variety of motives explain why NATO members opted for the aerial strategy in that case, such as the fact that United Nations Security Council Resolution 1973 excluded the possibility of a foreign occupation to protect civilians, but the main reason seems to reside in the simple fact that neither the US nor European countries were ready to pay the human costs of ground warfare.[21] The concern for casualties during the

[20] Krepinevich (1986, 164).

[21] Over the last few years, the practice of replacing troops with airpower has generated an intense theoretical debate among international relations scholars. The main point of contention is whether democracies are more likely to substitute manpower with firepower because of sensitivity to military casualties. For example, Caverley (2010, 2014) contends that democracies are more likely to overrely on firepower-intensive military doctrines, because they tend to be more palatable for the median voter. On the contrary, Gartzke (2001) finds no correlation between regime type and the use of capital-intensive weaponry and argues that states replace military manpower with firepower depending on the relative abundance of the latter in their economies. Accordingly, capital-abundant states use more weapons (tanks, ships, and aircrafts), whereas labor-abundant countries recruit more soldiers. Discussing the determinants of mechanization, Sechser and Saunders (2010) find that the mechanization levels used depend on the nature of the threat environment and

campaign against Libya was clearly expressed by a senior NATO officer, who declared, "The real challenge is public opinion in Europe and the nations' patience. They'd like the war to be over, and to have it done properly *with no allied casualties* or collateral damage to civilians."[22]

The increasing use of unmanned aerial vehicles (UAV) in various theaters like Iraq, Afghanistan, Yemen, Pakistan, and Syria is another example of the attempt to replace ground troops with airpower. Indeed, drone warfare offers a number of advantages compared with traditional air strikes: first, it guarantees the highest level of force protection, since the human pilot of the unmanned vehicle seats safely thousands of miles away from the theater of war. Second, and relatedly, relying on drones makes "search and rescue missions" no longer necessary when an aircraft is shot down by enemy air defense. Third, since, over the last few years, technological advancement has largely increased strike accuracy, today UAV can easily track their target, linger on it for several hours, and strike with great precision (i.e., their effectiveness has significantly increased). A recent empirical study, for example, has found that drone strikes in Pakistan have reduced (at least in the short term) the rate of terrorist attacks.[23] Furthermore, as Michael Horowitz reminds us in his chapter of this volume, drones are usually cheaper than their manned counterparts. Thus, it is not at all surprising that President Obama decided to heavily invest in the drone program and to use unmanned vehicles in several conflict areas to strike terrorists.

Protecting soldiers by replacing them with airpower is a noble objective in contrast to the past disregard for the deaths of combatants, but it must be emphasized that such a strategy has a variety of ethical, political, and war-fighting trade-offs. Concerning ethical issues, it is worth noting that a military approach based primarily on airpower increases the likelihood of collateral damage.[24] NATO's high-altitude bombing campaign in the Federal Republic of Yugoslavia, planned with the explicit goal of minimizing NATO casualties, is an exemplary case of

on the presence of internal threats, regardless of regime type. While the empirical record on the relationship between domestic politics and the preference for capital-intensive strategies is, therefore, mixed, scholars agree that such a military approach, if not militarily effective, can be considered today the favorite way of warfare for technologically advanced states.

[22] Cited in Schmitt and Sanger (2011), emphasis added.

[23] Johnston and Sarbahi (2016).

[24] Shaw (2002, 2005); Zehfuss (2011).

the trade-off between bombing and the possibility of discriminating between enemy combatants and civilians. In order to avoid military casualties, NATO members decided that they would not commit ground forces in Kosovo and opted instead for an air campaign.[25] Although minimizing harm to civilians was a real concern for Operation Allied Force, it nevertheless resulted in hundreds of unintended civilian casualties. As one scholar has recently observed in criticizing such a strategy: "Although militaries risk significantly higher casualties by deploying their troops rather than dropping precision bombs, this sacrifice is precisely what the logic of just war requires: that civilians not become more expendable than a country's armed forces."[26] From this point of view, Kosovo is a classic example of how the choice to protect troops puts the burden of war on enemy civilians.

This is especially true in nonconventional conflicts, where insurgents may use civilians (along with hostages)[27] as human shields. Although distance does not affect accuracy, the possibility of hitting only the right target is often jeopardized in a context where combatants and noncombatants are intermingled. For instance, during the 2014 Israel–Gaza conflict, an Israeli air strike hit a UN-operated school in Rafah, killing ten civilians. According to the Israeli government, the bombing was aimed at destroying Hamas's military installations, which had been deliberately placed by the Palestinian organization in such a building to inhibit enemy attacks. However, the Israeli shelling produced widespread international criticism and even condemnation from UNSG Ban Ki-moon.

Not only do airpower strategies cause negative ethical consequences, but they also imply war-fighting and political trade-offs. Some scholars,

[25] The absence of casualties during the war in Kosovo resulted from the fact that there was no ground war at all and NATO relied only on air strikes to deter the Serbians. Coker (2001, 12) argues that this would be a recurrent pattern, one already experienced by the US during the First Gulf War, where Americans expected 10,000 casualties and lost just 270 (that is, "only 40% of those who might have died had they stayed in the US"). Similarly, according to Record (2002, 10–11), Kosovo produced the "elevation of force protection above mission accomplishment."

[26] Carpenter (2011). For a study on civilian victimization in war, see Downes (2008).

[27] During the campaign in Pakistan in January 2015, for example, a US drone strike accidentally killed two Western hostages held in a compound by al-Qaeda, forcing President Obama to publicly apologize for the unintended victims.

for example, have noted that relying on drones may raise problems of accountability, legitimacy, and proportionality.[28] Additionally, while airpower strategies may not produce negative military outcomes in conventional theaters (where there is a clear-cut distinction between combatants and noncombatants),[29] in irregular warfare they may be highly counterproductive, because they can alienate local support and fuel the insurgency, bringing new recruits to rebel organizations. Over-relying on airpower means not only a reduced ability to distinguish between combatants and noncombatants, but also decreased capacity in winning the peace, which, according to Clausewitz, is the very goal of strategy. Indeed, irregular wars are often fought over the distribution of domestic political power and the belligerents' goal is usually to build a widespread consensus to rule over a contested territory. Therefore, attacks that kill civilians undermine popular support for the perpetrator. For example, a recent study by Kocher, Pepinsky, and Kalyvas shows that US aerial strategy during the Vietnam War systematically shifted popular support in favor of the Viet Cong.[30] Others suggest that not only was al-Qaeda in Iraq indiscriminate bombing against Shia mosques ineffective, but it also eventually resulted in its defeat.[31] Actually, bombing from a safe distance causing civilian casualties may bring new recruits to the insurgency groups, by showing to its domestic audience that the attacker is evil and must be resisted.

Moreover, aerial bombing is deliberately aimed at killing, not capturing rebels. This implies that relying on air strikes and drones may produce intelligence starvation, as combatants die without revealing important information. As a recent RAND commentary put it, "A drone attack destroys the critical intelligence that is needed to ensure that the tactical strike can be converted to strategic advantage."[32]

[28] See Brunstetter and Braun (2011); Boyle (2013); Peron (2014); Buchanan and Keohane (2015). This seems particularly true for the so-called signature strikes that allow a remote pilot to target individuals if some indicators of terrorist activity (such as gathering of people or traveling in convoys) are present, i.e., without the necessity of collecting further evidence about their actual identity.

[29] According to some analysts (e.g., Pape 1990; Horowitz and Reiter 2001), the adoption of such strategies is not problematic in conventional disputes, provided that the assailant targets military vulnerabilities only. Pape (2004) argues that airpower is highly effective only in combination with ground operations.

[30] Kocher et al. (2012).

[31] Ricks (2009); Phillips (2009).

[32] Robinson (2013).

Lastly, adopting a strategy based on airpower may limit military effectiveness insofar it excludes the possibility of controlling the territory, which is an important objective in irregular warfare.

Relying on airpower in irregular conflicts means incurring significant trade-offs, as mounting civilian casualties shift loyalty away from the attackers and may even contribute to strengthening the enemy. Although this seems to apply also to conventional wars,[33] in irregular fighting the problem is far more acute, since ground troops are usually required to provide security to the population and protect civilians from insurgents' attacks. In many conflicts where Western states have recently intervened, from Iraq to Afghanistan and Libya, either the absence or the small number of international ground troops meant the outbreak of ethnic, sectarian violence, and bloody intercommunal civil war. In sum, airpower is highly efficient in destroying things, crushing enemies, and removing political regimes, but it seems less than adequate when it comes to turning victory into a stable political outcome.

Deployment without Employment: Training, Compounds, and Indigenization

At the 2008 NATO Summit in Bucharest, Germany was criticized by some allies for having set high levels of force protection for its troops in Afghanistan. Reportedly, Bundeswehr troops used to "patrol only in armored personnel carriers" and did "not leave their bases at night."[34] To be sure, in the past, a number of other countries deployed their combatants without truly employing them (e.g., the South Koreans during the Vietnam War). Even in Iraq, when sectarian strife between Sunni and Shia groups erupted in December 2006, US Marines were still in their fortified compounds, and many commanders seemed more concerned with "protect[ing] their soldiers at all costs [rather] than with winning."[35]

When military personnel are deployed but not employed, force protection is set at a very high level, as military troops neither engage with the enemy nor do they approach it. Needless to say, when military troops stay in their compounds, high levels of force protection mean low levels of military effectiveness, and this generates some important

[33] Pape (1990). [34] Morelli and Belkin (2009, 10). [35] Ricks (2009, 34).

political and military implications. On the one hand, the conflict may result in a stalemate, or, in the worst scenario, irregular combatants may exploit the situation to spread their control over the country. Over time, this may erode public support for war, as citizens are likely to complain about conflict duration and financial costs. On the other, in lack of any successful result, political leaders may decide to exit the war short of victory and this could pave the way to new security threats. President Obama's decision to withdraw American troops from Iraq within December 2011, for example, may have unintendedly contributed to the spread of the Islamic State in the region.[36]

A second military practice that falls into this category of force protection is a heavy reliance on local forces. The policy of using indigenous forces has been termed in a variety of ways according to the geographical context. It was used in Vietnam (Vietnamization) and, more recently, in Iraq (Iraqization) and Afghanistan (Afghanization). Subcontracting fighting to local militias has now become a common practice for Western states, which prefer to engage in training missions, such as Italy does with Carabinieri in Iraq, Afghanistan and other post-conflict societies.

Local forces are valuable not only because they are familiar with local culture, as often argued,[37] but also because they sustain a suffering which Western societies are not ready to bear. In a comment to General Frank's decision to employ the Northern Alliance against the Taliban, US general Michael DeLong was absolutely explicit in stressing this point: "from our perspective, it was still less than it would have cost to put even one US battalion on the ground. In that sense, it was a bargain – and more important, *it would help keep thousands of US soldiers out of harm's way*."[38] From this point of view, the way the war in Libya was fought (unlike its outcome) seems the "ideal" war for casualty-sensitive countries: bombing from a safe distance and outsourcing the fight on the ground to indigenous forces.

Relying on indigenous forces might seem nothing new since European colonial powers always had their colonial troops, but today these native forces are not integrated in Western armies as they were in the past and, therefore, they are more difficult to control and to employ effectively. Actually, in relation to war fighting, the policy of relying

[36] Ralph and Souter (2015). [37] Cassidy (2006).
[38] Cited in Buley (2008, 115), emphasis added.

on indigenous troops is characterized by a variety of significant trade-offs. First, although local forces are usually advised, equipped, and trained by American and European troops, they retain their own organizations, agendas, and interests, which are often quite divergent from ours.[39] Moreover, relying on local militias may also produce new security threats: using indigenous forces implies not only training them, but also providing weapons, equipment, and ammunitions. And the more trained, armed, and equipped these local militias are, the greater the threat is if they decide to revolt against their former supporters. After all, backing the *mujaheddin* in Afghanistan against the Soviet Union contributed to a variety of unintended and adverse outcomes, such as the Afghan civil war, the Taliban regime, and the rise of al-Qaeda.[40]

Second, poor discipline, infiltration, and corruption among local troops affect the extent and the quality of their performance. On this point, Stephen Biddle found that native forces in Afghanistan performed very poorly in close combat and that skilled troops are essential to exploit the advantages of precision airpower.[41] Furthermore, the lack of professionalization of indigenous forces may also result in human rights abuses, rapes, and other war crimes, prompting local disaffection to an even higher degree. Third and lastly, local militias and police forces may easily become the target of attacks. Insurgents are in competition with the government, and by selectively attacking local militias they may discourage recruitment. Al-Qaeda in Iraq, for example, routinely attacked recruitment sites in Baghdad with suicide bombers, killing several Iraqi citizens willing to join the army, with the clear intention of intimidating the locals and discouraging enlistment.[42]

While relying on local forces might save soldiers' lives, it must be emphasized that they are not always the ideal instrument to achieve the political goals which Western states fight for. Actually, the use of indigenous armed groups may negatively impact military effectiveness (because of their poor skills) and even create new security threats.

[39] Byman (2006).

[40] The recent debate about arming rebels in Syria against the Assad regime was framed by critics around this argument.

[41] Biddle (2006a). For a different line of argument, see Andres et al. (2006), who argue that the policy of indigenization in Afghanistan (what they call the "Afghan model") may apply to other conflicts, but only in combination with other military practices (i.e., the deployment of special operation forces and airpower).

[42] Kydd and Walter (2006).

Of course, as Stephen Biddle put it, the nature of the war can also make a significant difference.[43] In particular, indigenous troops appear better fit for people's wars, such as the final phase of the Vietnam conflict, than for a communal civil war, like in today's Iraq. As Biddle (2006b) argued, communal civil conflicts are different from "Maoist," people's wars: the latter are characterized by the "struggle for good governance between a class-based insurgency claiming to represent the interests of the oppressed public," while the former "feature opposing subnational groups divided along ethnic or sectarian lines." In this light, supporting indigenous troops may raise other groups' fear, thus aggravating the security dilemma and fueling violence in the country.

Ground Warfare Force Protection: Micromeasures for Casualty Minimization

Beyond airpower and indigenous forces, there is a third type of military practices employed to minimize friendly casualties when sending boots on the ground is inevitable: mechanized land warfare and a variety of microtactical measures designed to reduce soldiers' exposure to enemy fire. While in the past these micromeasures mainly dealt with the defense of military installations and bases, with the wars in Kosovo, Iraq, and Afghanistan adopting these force protection measures has become a priority in combat operations too. These practices might concern relevant tactical devises regarding rules of engagement (e.g., checkpoint protocols, targeting, urban combat), decisions on *weaponeering*, and the level at which the threshold for using force is set.[44] As mentioned above, all these practices can be termed *ground warfare force protection.*

In order to show how these measures impact on military effectiveness in irregular warfare, three different types of practices have been considered here, which imply different degrees of exposure to enemy fire. A first option is to rely on heavy mechanized warfare, as the Russians did, for example, during the early stages of the first Chechen war in 1995. In conventional warfare, heavy armored vehicles allow both a great deal of effectiveness and high levels of force protection, since they can be easily moved from one front to another and perform particularly well in the open field. In the past, tank deployment was also

[43] Biddle (2006b). [44] T. W. Smith (2008).

used by autocratic governments to occupy dissident cities (like during the Prague Spring in 1956) and to repress urban riots (like during the Tiananmen Square protests in May 1989). However, in irregular conflicts, they can be highly ineffective, since speed and maneuver can be limited by mountainous and rough terrain, as occurred, for example, to the French armed forces in Indochina. Moreover, in an urban environment, armored vehicles may represent an easy target for ambush, if rebels are equipped with antitank weapons.

Overall, there are at least two trade-offs that emerge from the use of highly armored vehicles to face irregular enemies. On the one hand, there is still the problem of discriminating between foes and the rest of the population. As a variety of scholars have shown,[45] mechanized warfare may generate similar effects as long-distance bombing increasing the difficulty of distinguishing between soldiers and civilians in a war zone. By employing heavy weaponry, this type of force protection reduces the ability of distinguishing between enemy combatants and noncombatants and, inevitably, causes collateral damage. On the other hand, there are intelligence problems.[46] Built for conventional warfare, heavy mechanized troops tend to be ineffective in irregular conflicts, because they increase the physical distance between soldiers and civilians and thus impede information gathering on who enemy combatants actually are, on where they hide, and where they place antitank devices. Additionally, as Lyall and Wilson put it, in mechanized warfare "it . . . becomes harder to recruit reliable collaborators among local populations, further compounding information starvation. The result is a counter-insurgent that fuels, rather than deters, insurgent recruitment."[47]

Opting for a lower degree of armoring (i.e., using light mechanized units) may solve some of the problems outlined above, but generates a variety of other negative outcomes, as the Americans experienced in recent conflicts. Until not long ago, US armed forces in Iraq and Afghanistan made extensive use of Humvees and other hybrid vehicles, both for patrolling and for actual fighting. Compared to heavier armored tanks, these multipurpose vehicles provide troops with more

45 Lyall and Wilson (2009); Zambernardi (2010); Levy (2010).
46 Lyall and Wilson (2009).
47 Lyall and Wilson (2009, 75–76). In his study on civilian defense forces, Peic (2014) finds that higher levels of mechanization in irregular warfare can be actually associated with successful COIN operations.

agility and mobility, and are better suited for irregular combat. Moreover, given that they are armored with hardened steel and bulletproof glass, they seem to assure high levels of force protection for military personnel. However, this kind of vehicle can be also highly ineffective to counter an irregular enemy, because – like in heavy mechanized warfare – collecting intelligence may be difficult for mounted soldiers. Worse still, as US Marines realized soon after the eruption of intercommunal violence in Iraq (i.e., between the end of the conventional phase in May 2003 and the Surge in early 2007), they may also prove ineffective to protect troops. Indeed, US casualties during this period (about half of all hostile deaths) were mainly due to improvised explosive devices or roadside bombs that hit American personnel during transfers and patrolling operations. On this point, it is worth noting that an overreliance on mounted operations meant, quite paradoxically, that just one device, if carefully placed by insurgents, could be very deadly for US military personnel. In this case, decreasing levels of force protection to gain mobility and maneuverability resulted in a far higher degree of exposure for US forces. That is why, by mid 2007, the Pentagon decided to employ also the new Mine Resistant Ambush Protected (MRAP) vehicles, which would prove safer.[48]

A third option in ground warfare is relying on dismounted operations, which, according to some military analysts, represent the essential core of every counterinsurgency effort.[49] Clearly enough, in order to implement these operations, force protection measures have to be significantly lowered, while soldiers must be trained to manage higher levels of exposure to enemy fire, meaning that they have to resist the temptation to overretaliate in the face of friendly casualties. Again, a close look at the war in Iraq may be instructive for understanding the trade-offs involved in these measures, in particular when conventional warfare turned into an irregular conflict. In 2004, as a response to mounting friendly casualties, US troops had increased the level of force protection by reducing the threshold at which coercive force could be used, as exemplified by the battle of Fallujah (November–December 2004) and by the Haditha incident (November 19, 2005), in which US Marines killed twenty-four people in response to a roadside bomb, of

[48] On this point, see Lamb et al. (2009).

[49] See, among others, Krepinevich (2005); Kilcullen (2006); Ricks (2009).

whom nineteen were harmless civilians who lived in the houses along the road. Although the tragedy of Haditha was not the direct consequence of official US military policy, the massacre was no coincidence, but rather the latest of a long series of incidents in which US troops overreacted in a situation of danger. In a way, Haditha epitomized the initial US approach to the insurgency in Iraq, which was fought with the primary goal of minimizing friendly casualties and on the tacit assumption that all Iraqis were a potential threat.[50]

Needless to say, when force protection measures are too much lowered to meet the requirements of irregular warfare, this may generate higher levels of military casualties, thus impacting on political support for war and prompting states to exit the conflicts short of victory. The experience of the US in Iraq is worth pondering at length.

In early 2007 President Bush announced a boost of about thirty thousand US forces in Iraq (i.e., the Surge) and the adoption of a new strategy to fight the mounting insurgency. This change of strategy envisaged the rejection of previous notions of force protection and surely produced an impact on how military operations were conducted. In particular, the deployment of additional troops to Iraq meant not only greater manpower, but it also brought more troops to carry out dismounted and ground operations. Under the command of General Petraeus, American armed forces adopted a COIN approach, shifting from "enemy-centric" counterterrorism to a "population-centric" strategy. The priority, according to Petraeus, was in the nonkinetic aspect of the conflict: the population needed to be secured, even if this meant giving up "search and destroy" operations. In other words, instead of concentrating the effort of US troops on training Iraqi forces, Petraeus decided that it was American troops who had to take primary responsibility for securing and stabilizing the territory, as not only were Iraqi troops and police forces not properly trained and equipped, but they were also part of the problem. Indeed, the Shi'a-dominated government was not a political neutral institution but a sectarian combatant in the mounting Iraqi civil war.[51] The Surge – it should be emphasized – was more about "how to use troops" than it was about numbers. It was organized through "slow, ambiguous operations that were built not only around technology but around human interactions."[52] The goal turned into "winning the people" through

[50] Kilcullen (2009, 124). [51] Kilcullen (2009, 126–127).
[52] Ricks (2009, 162–165).

a variety of measures that attempted to provide security to the Iraqi population, such as the establishment of numerous small residential outposts, an almost continuous dismounted patrolling of the streets, and, more generally, keeping boots on the ground. In so doing, the US adopted what David Kilcullen termed the residential approach: "living in your sector, in close proximity to the population rather than raiding into the area from remote, secure bases."[53] Then, once security was established in a particular area, economic resources were directed toward the gradual, short- and long-term well-being of the civilian population.

Not surprisingly, the new approach resulted in a higher number of US casualties. Indeed, the hostile death rate increased, with its monthly peak of 126 in May 2007. It should be emphasized that, over the next few months, this doctrinal change – along with the injection of additional troops and the so-called Anbar Awakening[54] – would actually prove successful in reducing both the number of friendly casualties[55] and the overall level of violence in Iraq, as the number of noncombatant fatalities declined "by 80 percent by 2008 and 90 percent by early 2009."[56] However, besides President Obama's determination to follow through on a campaign promise, a combination of cumulative casualties, length, and financial costs of conflict seems to have induced the American administration to withdraw from Iraq. In other words, the new approach succeeded in reducing violence and improving security in Iraq only temporarily. As it is tragically clear today, the decision to withdraw US and coalition troops in 2011 opened the gate for the Islamic State jihadists, who exploited the power vacuum created by the American exit from the country.

Conclusion

In this chapter, we have argued that one of the essential features of modern warfare – the protection of soldiers' lives as an objective in itself, regardless of strategic and tactical considerations – has a variety

[53] Kilcullen (2006, 105). [54] Biddle et al. (2012); Gentile (2013, 87–88).

[55] Indeed, after the early months of the Surge, the casualty rate for US troops declined (Goldberg 2010, 2014). To be more specific, while from 2004 to mid 2007, the US military was suffering a monthly toll of almost seventy-four dead, in the last three months of 2007, the toll declined to thirty-three. A similar trend was recorded for the number of WIAs (wounded in action).

[56] Campbell et al. (2009, 22).

of trade-offs in terms of military effectiveness. These trade-offs emerge especially in irregular warfare and when states are committed to avoid a policy of barbarism (i.e., massive, indiscriminate killing of civilians). After all, ruthless political leaders facing an insurgency may choose to rely on brutality to crack down on insurgents along with innocent people. Russian armed forces during the Second Chechen War, for example, used indiscriminate attacks (such as artillery strikes and massive shelling of civilian areas) to suppress the Chechen insurgency.[57] Likewise, in 2009 the Tamil insurgency in Sri Lanka was crushed by the government through brutal coercive methods.[58]

By contrast, states that are committed to discriminating between combatants and noncombatants will necessarily face a number of trade-offs, whose nature will depend on the level of force protection measures.[59] First, airpower strategies generate both ethical and strategic trade-offs, as bombing from a safe distance may result in reduced ability to discriminate combatants from the bulk of the population, thus undermining the locals' support and bringing new recruits to the insurgency groups. Second, relying on indigenous forces entails problems of troop control and training, and may even generate new security threats. Third, mechanized land warfare may produce information problems, while mounting friendly casualties may induce troops to overretaliate against civilians. Reducing force protection measures and protecting the population proved effective in Iraq after the Surge, and it resulted in decreased levels of both civilian and friendly casualties, though this success was temporary and due also to other intervening factors, such as the Anbar Awakening. This primarily means that population-centric COIN should not be assumed as a panacea for irregular conflicts, because dismounted troops must be trained to

[57] See Lyall (2009). In their study on public perception of the Second Chechen war, Theodore Gerber and Sarah Mendelson (2008) find that Russian society – though casualty sensitive – was not that concerned about human rights.

[58] Hashim (2013).

[59] The revised edition of FM 3-24, issued in May 2014, seems to have caught this point when it states that "different capabilities provide different choices that offer different costs and risks" and that "commanders inform political leaders about the prospects for victory and the different costs and risks of various options, and political leaders weigh these costs and risks against their importance to U.S. national interests." See https://fas.org/irp/doddir/army/fm3-24.pdf.

manage such a change in military strategy and their societies must be able to tolerate mounting casualties.

Additionally, since public support for war stems from people's attitudes, the way political elites frame military interventions may matter as well.[60] On the one hand, wars of national liberation and nationalist disputes may be framed as "wars of necessity," encouraging soldiers to fight to the death and inducing the public to tolerate casualties. On the other, foreign interventions in conflict areas are most likely to be viewed as "wars of choice."[61] Public support for war may be more volatile in this case. In short, the nature of war may have important implications not only for war fighting,[62] but also for domestic support.

The trade-offs between force protection and military effectiveness in irregular conflicts may also have significant implications for future warfare, which is likely to become increasingly unconventional over the next few years. While securing your own forces may be an efficient strategy to minimize friendly casualties and maintain public support for war, relying on force protection measures may induce generals and commanders to adopt ineffective military doctrines, which make victory more unlikely.[63] Thus, in irregular conflicts troops should restrain from engaging in mechanized operations and focus instead on low-tech tactics, such as patrolling the streets, establishing outposts, moving into the population and being "out there." This implication leads to a basic question: how can strategists reconcile the trade-off between the increasing sensitivity to friendly casualties and the need to disregard force protection measures in order to protect civilians in irregular conflicts? At present this question does not have an easy and straightforward solution.[64] In other words, it is a dilemma that political and military leaders are likely to face, over and over again, in the coming decades.

[60] Berinsky (2007, 2009) [61] Haas (2009). [62] Biddle (2006b).

[63] On this point, Virginia Page Fortna (2009) argues that draws have become more common over time, both in inter- and in intrastate conflicts. It is not the goal of this chapter to investigate the relationship between casualty sensitivity and war outcome. However, we acknowledge that both trends (increasing levels of force protection and more wars resulting in draws) may be related, especially in civil conflict.

[64] This may be especially true if we consider the close link between the way of fighting and the social, cultural, and political features of the country that adopts it. See Andreatta (forthcoming).

References

Andreatta, Filippo. 2012. The Regularity of Irregular War: Counterinsurgency and Its Implications. *Finmeccanica Working Papers*, 1–44.

Forthcoming. Military Power and the Art of War. In *From the Polis to Absolute States*. Trento: FBK Press.

Andres, Richard B., Craig Wills, and Thomas E. Griffith, Jr. 2006. Winning with Allies: The Strategic Value of the Afghan Model. *International Security* 30(3): 124–160.

Arreguín-Toft, Ivan. 2005. *How the Weak Win Wars: A Theory of Asymmetric Conflict*. New York: Cambridge University Press.

Berinsky, Adam J. 2007. Assuming the Costs of War: Events, Elites, and American Public Support for Military Conflict. *Journal of Politics* 69(4): 975–997.

2009. *In Time of War: Understanding American Public Opinion from World War II to Iraq*. Chicago: University of Chicago Press.

Biddle, Stephen D. 2004. *Military Power: Explaining Victory and Defeat in Modern Battle*. Princeton, NJ: Princeton University Press.

2006a. Allies, Airpower, and Modern Warfare: The Afghan Model in Afghanistan and Iraq. *International Security* 30(3): 161–176.

2006b. Seeing Baghdad, Thinking Saigon. *Foreign Affairs* 85(2):2–14.

Biddle, Stephen D., Jeffrey A. Friedman, and Jacob N. Shapiro. 2012. Testing the Surge: Why Did Violence Decline in Iraq in 2007? *International Security* 37(1): 7–40.

Boyle, Michael J. 2013. The Costs and Consequences of Drone Warfare. *International Affairs* 89(1): 1–29.

Brunstetter, Daniel, and Megan Braun. 2011. The Implications of Drones on the Just War Tradition. *Ethics and International Affairs* 25(3): 337–358.

Buchanan, Allen, and Robert O. Keohane. 2015. Toward a Drone Accountability Regime. *Ethics and International Affairs* 29(1): 15–37.

Buley, Ben. 2008. *The New American Way of War: Military Culture and the Political Utility of Force*. London: Routledge.

Byman, Daniel L. 2006. Friends Like These: Counterinsurgency and the War on Terrorism. *International Security* 31(2): 79–115.

Campbell, Jason, Michael O'Hanlon, and Jeremy Shapiro. 2009. How to Measure the War. *Policy Review* 157(October/November): 15–30.

Carpenter, Charli. 2011. Fighting the Laws of War. *Foreign Affairs* 90(2): 146–152.

Cassidy, Robert M. 2006. The Long Small War: Indigenous Forces for Counterinsurgency. *Parameters* 36: 47–62.

Caverley, Jonathan D. 2010. The Myth of Military Myopia: Democracy, Small Wars, and Vietnam. *International Security* 34(3): 119–157.

2014. *Democratic Militarism: Voting, Wealth, and War*. Cambridge: Cambridge University Press.
Clausewitz, Carl, von. 1993. *On War*. London: Everyman's Library.
Coker, Christopher D. 2001. *Humane Warfare: The New Ethic of Post Modern War*. London: Routledge.
Cornish, Paul. 2003. Myth and Reality: US and UK Approaches to Casualty Aversion and Force Protection. *Defense Studies* 3(2): 121–128.
Downes, Alexander B. 2008. *Targeting Civilians in War*. Ithaca, NY: Cornell University Press.
Fortna, Virginia Page. 2009. Where Have All the Victories Gone? Peacekeeping and War Outcomes. Paper presented at the annual meeting of the American Political Science Association, Toronto, September 6. www.columbia.edu/~vpf4/victories%20Sept%202009.pdf.
Gartzke, Eric. 2001. Democracy and the Preparation for War: Does Regime Type Affect States' Anticipation of Casualties? *International Studies Quarterly* 45(3): 467–484.
Gelpi, Christopher. 2006. The Costs of War: How Many Casualties Will Americans Tolerate? *Foreign Affairs* 85(1): 139–144.
Gelpi, Christopher, and Peter Feaver. 2004. *Choosing Your Battles: American Civil-Military Relations and the Use of Force*. Princeton, NJ: Princeton University Press.
Gelpi, Christopher, Peter Feaver, and Jason Reifler. 2005. Success Matters: Casualty Sensitivity and the War in Iraq. *International Security* 30(3): 7–46.
2009. *Paying the Human Costs of War: American Public Opinion and Casualties in Military Conflicts*. Princeton, NJ: Princeton University Press.
Gentile, Gian. 2013. *Wrong Turn: America's Deadly Embrace of Counterinsurgency*. New York: New Press.
Gerber, Theodore P., and Sarah E. Mendelson. 2008. Casualty Sensitivity in a Post-Soviet Context: Russian Views of the Second Chechen War, 2001–2004. *Political Science Quarterly* 123(1): 39–68.
Goldberg, Matthew S. 2010. Death and Injury Rates of US Military Personnel in Iraq. *Military Medicine* 175(4): 220–226.
2014. Updated Death and Injury Rates of US Military Personnel during the Conflicts in Iraq and Afghanistan. Working Paper 2014–08, Congressional Budget Office, Washington, DC. www.cbo.gov/publication/49837.
Grauer, Ryan, and Michael C. Horowitz. 2012. What Determines Military Victory? Testing the Modern System. *Journal of Conflict Resolution* 21(1): 83–112.

Gray, Chris H. 2003. Posthuman Soldiers in Postmodern War. *Body and Society* 9(4): 215–226.

Griffin, Christopher. 2010. Major Combat Operations and Counterinsurgency Warfare: Plan Challe in Algeria, 1959–1960. *Security Studies* 19(3): 555–589.

Haas, Richard N. 2009. *War of Necessity. War of Choice. A Memoir of Two Iraq Wars*. London: Simon and Schuster.

Hanson, Victor D. 1989. *The Western Way of War: Infantry Battle in Classical Greece*. New York: Alfred A. Knopf.

Hashim, Ahmed. 2013. *When Counterinsurgency Wins: Sri Lanka's Defeat of the Tamil Tigers*. Philadelphia: University of Pennsylvania Press.

Hoffman, Frank. 2007. *Conflict in the 21st Century: The Rise of Hybrid Wars*. Arlington, VA: Potomac Institute for Policy Studies.

Horowitz, Michael C., and Dan Reiter. 2001. When Does Aerial Bombing Work? Quantitative Empirical Tests, 1917–1999. *Journal of Conflict Resolution* 45(2): 147–173.

Horowitz, Michael C., Erin M. Simpson, and Allan C. Stam. 2011. Domestic Institutions and Wartime Casualties. *International Studies Quarterly* 55(4): 909–936.

Johnston, Patrick B., and Anoop K. Sarbahi. 2016. The Impact of US Drone Strikes on Terrorism in Pakistan and Afghanistan. *International Studies Quarterly* 60(2): 203–219.

Kaldor, Mary. 1998. *New and Old Wars: Organized Violence in a Globalized Era*. Stanford, CA: Stanford University Press.

Keegan, John, and Richard Holmes. 1985. *Soldiers: A History of Men in Battle*. London: Hamish Hamilton.

Kilcullen, David. 2006. Twenty-Eight Articles: Fundamentals of Company-Level Counterinsurgency. *Military Review* 86(May–June): 103–108.

2009. *The Accidental Guerrilla: Fighting Small Wars in the Midst of a Big One*. Oxford: Oxford University Press.

Kocher, Matthew A., Thomas B. Pepinsky, and Stathis N. Kalyvas. 2012. Aerial Bombing and Counterinsurgency in the Vietnam War. *American Journal of Political Science* 55(2): 201–218.

Krepinevich, Andrew F. 1986. *The Army and Vietnam*. Baltimore: Johns Hopkins University Press.

2005. How to Win in Iraq. *Foreign Affairs* 84(5): 87–104.

Kydd, Andrew H., and Barbara F. Walter. 2006. The Strategies of Terrorism. *International Security* 31(1): 49–80.

Lamb, Christopher J., Matthew Schmidt, and Berit G. Fitzsimmons. 2009. MRAPS, Irregular Warfare, and Pentagon Reform. Occasional Paper 6,

Institute for National Strategic Studies, National Defense University. http://usacac.army.mil/cac2/cgsc/sams/media/MRAPs.pdf.

Levy, Yagil. 2009a. An Unbearable Price: War Casualties and Warring Democracies. *International Journal of Politics, Culture, and Society* 49(6): 69–82.

2009b. From Republican to Market Control over the Armed Forces: A Conceptual Framework. *Polity* 41(1): 1–29.

2010. The Tradeoff between Force and Casualties: Israel's Wars in Gaza, 1987–2009. *Conflict Management and Peace Science* 27(4): 386–405.

2012. *Israel's Death Hierarchy: Casualty Aversion in a Militarized Democracy*. New York: New York University Press.

Lyall, Jason. 2009. Does Indiscriminate Violence Incite Insurgent Attacks? Evidence from Chechnya. *Journal of Conflict Resolution* 53(3): 331–362.

Lyall, Jason, and Isaiah Wilson. 2009. Rage against the Machines: Explaining Outcomes in Counterinsurgency Wars. *International Organization* 63(1): 67–106.

Mearsheimer, John J. 1983. *Conventional Deterrence*. Ithaca, NY: Cornell University Press.

Millett, Allan R., Williamson Murray, and Kenneth H. Watman. 1988. The Effectiveness of Military Organizations. In Allan R. Millett and Williamson Murray, eds., *Military Effectiveness*, vol. 1, 1–30. Boston: Allen and Unwin.

Morelli, Vincent, and Paul Belkin. 2009. *NATO in Afghanistan: A Test of the Transatlantic Alliance*. CRS Report for Congress. https://fas.org/sgp/crs/row/RL33627.pdf.

Mueller, John. 1989. *Retreat from Doomsday: The Obsolescence of Major War*. New York: Basic Books.

Münkler, Herfried. 2005. *The New Wars*. Cambridge: Polity Press.

Pape, Robert A. 1990. Coercive Air Power in the Vietnam War. *International Security* 15(2): 103–146.

2004. The True Worth of Air Power. *Foreign Affairs* 83(2): 116–130.

Peic, Goran. 2014. Civilian Defense Forces, State Capacity, and Government Victory in Counterinsurgency Wars. *Studies in Conflict and Terrorism* 37(2): 162–184.

Peron, Alcides E. R. 2014. The "Surgical" Legitimacy of Drone Strikes? Issues of Sovereignty and Human Rights in the Use of Unmanned Aerial Systems in Pakistan. *Journal of Strategic Security* 7(4): 81–93.

Phillips, Andrew. 2009. How al Qaeda Lost Iraq. *Australian Journal of International Affairs* 63(1): 64–84.

Pinker, Steven. 2011. *The Better Angels of Our Nature: Why Violence Has Declined.* New York: Viking Press.

Ralph, Jason, and Jamer Souter. 2015. A Special Responsibility to Protect: The UK, Australia and the Rise of Islamic State. *International Affairs* 91(4): 709–723.

Record, Jeffrey. 2002. Collapsed Countries, Casualty Dread, and the New American Way of War. *Parameters* 32(2): 4–23.

Reiter, Dan. 2007. Nationalism and Military Effectiveness: Post-Meiji Japan. In Risa A. Brooks and Elizabeth A. Stanley, eds., *Creating Military Power*, 27–54. Stanford, CA: Stanford University Press.

Ricks, Thomas E. 2009. *The Gamble: General Petraeus and the Untold Story of the American Surge in Iraq.* New York: Penguin.

Robinson, Linda. 2013. The Downside of Drones. Rand Corporation. www.rand.org/blog/2013/11/the-downside-of-drones.html.

Rothenberg, Gunther. 2000. *The Napoleonic Wars.* London: Cassell.

Schmitt, Eric, and David E. Sanger. 2011. As Goal Shifts in Libya, Time Constrains NATO. *New York Times*, May 26. www.nytimes.com/2011/05/27/world/africa/27policy.html?_r=0.

Sechser, Todd S., and Elisabeth N. Saunders. 2010. The Army You Have: The Determinants of Military Mechanization, 1979–2001. *International Studies Quarterly* 54(2): 481–511.

Shaw, Martin. 2002. Risk-Transfer Militarism, Small Massacres and the Historic Legitimacy of War. *International Relations* 16(3): 343–359.

2005. *The New Western Way of War.* Cambridge: Polity Press.

Showalter, Dennis, and William J. Astore. 2007. *Soldiers' Lives through History: The Early Modern Period.* New York: Greenwood Press.

Smith, Hugh. 2005. What Costs Will Democracies Bear? A Review of Popular Theories of Casualty Aversion. *Armed Forces and Society* 31(4): 487–512.

Smith, Rupert. 2008. *The Utility of Force: The Art of War in the Modern World.* New York: Vintage Books.

Smith, Thomas W. 2008. Protecting Civilians . . . or Soldiers? Humanitarian Law and the Economy of Risk in Iraq. *International Studies Perspectives* 9(2): 144–164.

Strachan, Hew. 2010. Strategy or Alibi? Obama, McChrystal and the Operational Level of War. *Survival* 52(5): 157–182.

Summerville, Christopher. 2005 (1974). Foreword. In H. C. B. Rogers, ed., *Napoleon's Army.* Barnsley, UK: Pen and Sword.

Valentino, Benjamin A., Paul K. Huth, and Sarah E. Croco. 2010. Bear Any Burden? How Democracies Minimize the Costs of War. *The Journal of Politics* 72(2): 528–544.

Van Creveld, Martin. 1991. *The Transformation of War*. New York: Free Press.

Walzer, Michael. 2006 (1977). *Just and Unjust Wars: A Moral Argument with Historical Illustrations*. New York: Basic Books.

Zambernardi, Lorenzo. 2010. Counterinsurgency's Impossible Trilemma. *Washington Quarterly* 33(3): 21–34.

2017. Life and Death in the Western Way of War, from the Modern Period to the Present. Unpublished manuscript.

Zehfuss, Maja. 2011. Targeting: Precision and the Production of Ethics. *European Journal of International Relations* 17(3): 543–566.

3 War Finance and Military Effectiveness

ROSELLA CAPPELLA ZIELINSKI

Any flowchart of military effectiveness begins with resources and their procurement, while flowcharts of readiness and force structure end with the amount of money to be requested.[1]

Money as a critical and necessary component of warfare is unquestioned. Yet, seldom discussed is the relationship between how states acquire it, the costs of acquisition, and how those costs affect the war effort. Moreover, the relationship between money and war is tenuous. While the notion that more money is better looms large, various resource poor states (e.g., Turkey during the Russo–Turkish War of 1877–1878;[2] Mexico in its war of liberation against France 1862–1867[3]) have gone on to fight much longer and more victoriously than anticipated.[4] Thus, it is no surprise that some studies find no correlation between a state's wealth and war outcome and others do.[5]

This chapter contributes to our understanding of the relationship between money and war by examining not the amount of funds a state can procure but the manner in which the state does so. Through the study of war finance, we gain insight into the costs and trade-offs leaders face when pursuing a conflict. Specifically, the manner in which a state finances a war affects public support for the war, security threats beyond the war, and war fighting.

I build upon the war finance literature and previous research highlighting the role of financial power in conflict. I argue that each method of war finance – taxation, debt, and sovereign-to-sovereign loans – has implications for the relationship between the state and the resource

[1] For example, see Brooks (2007, 4); Tellis et al. (2000, 137); Harrison (2014, 61).

[2] Turkey, characterized with extreme poverty at the time, was able to fight much longer than anticipated, to the surprise of British military observers. See Hozier (2005, 604–605).

[3] For a discussion of Mexican fiscal capacity, see Centeno (1997).

[4] Arreguín-Toft (2005, 4); Brooks (2003, 154).

[5] Biddle (2004, 21–23). Conversely, Stam (1996) and Beckley (2010) find national wealth correlates with victory.

extraction source, and, consequently, the war effort. When wars decline in popularity and are financed by taxation, citizens express their dissatisfaction by pressuring lawmakers to constrain the military budget. A newly constrained budget, coupled with the need to fund a continuing war effort, forces not only a guns-versus-butter budgetary trade-off but also a guns-versus-guns military spending trade-off. This guns-versus-guns trade-off results in the reappropriation of funds away from other national security needs toward the war effort.

Alternatively when procuring funds abroad, the state becomes indebted to foreign creditors. When those foreign creditors are states, they can use their newfound leverage, the threat to withhold funds, to influence the belligerent's policy choices. In essence, the creditor state forces the belligerent to modify its war aims, including early war termination.

I explore the relationship between war finance and military effectiveness through two exploratory cases using extreme case selection.[6] To locate extreme cases, I draw upon the Confronting the Cost of War data set that documents the financing of interstate wars from 1823 to 2003.[7] In regards to taxation, I explore the effects of President Truman's pay-as-you-go financing of the Korean War.[8] In the early months of the war, strong public support for the war overcame any aversion to financial sacrifice. Once public support for the war declined, citizens felt that they sacrificed enough. As a result, an upper bound was implemented on the amount of resources that could be mobilized for the war effort. To cope with a capped military budget the services reappropriated funds dedicated to meeting security goals beyond the war.

In regards to sovereign-to-sovereign loans, I explore the US loan to Israel during the 1973 Arab–Israeli War.[9] Israel needed a sovereign-to-sovereign loan in the form of military equipment from the US in order

[6] Seawright and Gerring (2008, 301); Gerring (2017, 55–57).

[7] Cappella Zielinski (2016, Chapter 6). The majority of states engage in multiple methods of war finance to finance a war (Cappella Zielinski 2016, 110). Hence, an extreme case selection is also attractive, as it allows the variable of interest, e.g., war finance method, to be isolated.

[8] While states may implement a war tax (e.g., Flores-Macías and Kreps 2013) and attempt to pay for a significant cost of the war via taxation, taxation typically comprises a small percentage of a state's war finance strategy (Cappella Zielinski 2016, 108–109).

[9] For a list of states that have extended and received a sovereign-to-sovereign wartime loan, see Cappella Zielinski and Poast (2015).

to continue its war effort. In turn, Israel became dependent on continued American supply. This newfound dependence on the US allowed the US to influence Israeli war fighting and settlement negotiation, limiting Israel's pursuit of Egyptian forces in order to implement a cease-fire.

Taken together, these findings suggest diffuse forms of war finance, i.e., debt, are the most effective. Whereas taxation results in a guns-versus-guns trade-off and sovereign-to-sovereign loans reduce state autonomy, debt allows leaders to reduce the perceived costs of war citizens bear and, thus, avoid various trade-offs. This finding is consistent with and adds to the growing body of literature that suggests borrowing is associated with mobilization advantages and preservation of leadership tenure during conflict.[10]

This chapter makes several contributions to the understanding of military effectiveness. First, it develops a novel model of the relationship between war finance and warfare. Second, it links this model with early scholarship on economic sources of military power.[11] Recent neoclassical realist literature emphasizes that primacy in the international system is subject to domestic political institutional constraints. Extractive and mobilization capacity is a crucial intervening variable between systemic imperatives and the foreign and defense policies states pursue.[12]

Third, this chapter provides both a temporal and across-state variation model to understand military effectiveness. Regime type, culture, nationalism, ethnicity, social structure, and identity may not vary during the conflict. These variables seek to explain military effectiveness advantages across states. How a state finances war varies across states and during the conflict. As Richard Betts states, "Immediate capacity tells us little about results. Unless it can ensure that the battle is decided on the first day, a force that could fight spectacularly for one day but would collapse the second is no more ready in any meaningful sense

[10] For mobilization advantages, see Shea (2014) and DiGiuseppe (2015b). For preservation of leadership tenure during conflict, see DiGiuseppe (2015a) and DiGiuseppe and Shea (2016).

[11] Early studies on economic sources of military power (Organski and Kugler 1980; Kugler and Domke 1986) argued that the ability to extract resources from society, particularly tax revenue, was essential for projecting military power. See also Gilpin (1981); Knorr (1975); Lamborn (1983).

[12] Lobell et al. (2009).

than is one that could not fight as well but could fight longer."[13] Like war, war finance is an evolving process. As a conflict unfolds, various policy makers within a state make decisions regarding the amount of resources devoted to it. Hence, where these resources come from may affect the war effort.

This chapter proceeds as follows. First, I review the literature addressing the relationship between resource mobilization and war. Second, I present my theoretical framework. I discuss the various ways wars can be financed and how said variation affects public support for war, security trade-offs, and war fighting. Third, I illustrate the effects of war finance by looking at two cases of limited conventional interstate wars in the post–World War II era: Truman's financing of the Korean War and US loans to Israel during the 1973 Arab–Israeli War. I conclude with a discussion of future research concerning war finance to military effectiveness.

The Study of Money and War

The existing literature that lies at the nexus of marshaling resources for war and war fighting provides little insight into military effectiveness. The war finance literature emphasizes how states finance their wars and overlooks the effects of war finance on the war effort. These studies seek to explain the war finance method chosen by understanding the costs to leaders vis-à-vis their tenure and their desire to advance domestic policy objectives though wartime mobilization.[14] The effects of the method chosen on the war is either unaddressed or subsumed under the notion "the more resources the better."

The literature that seeks to understand the relationship between money and war outcome presumes the more money, the better; thus, those states that can procure more resources are more likely to be successful in war.[15] This body of research overwhelmingly emphasizes the

[13] Betts (1995, 37).

[14] Levi (1988); Flores-Macías and Kreps (2013); Barnett (1992); Carter and Palmer (2015); Cappella Zielinski (2016).

[15] The presumption of more money is better can be seen in early works, such as Organski and Kugler (1980) and Kugler and Domke (1986). This presumption often manifests in the use of gross indicators of state wealth – GDP, GDP per capital, COW's CINC – to predict war outcome (e.g., Lake 1992; Stam 1996). It should be noted that Reiter and Stam (2002, 120) find no correlation between amount of matériel an army actually has in battle and battle outcomes.

role of debt and regime type. In regards to debt, states that can float more debt are more likely to be victorious.[16] Schultz and Weingast find democracies are best able to secure said debt and, thus, are more likely to win enduring rivalries.[17] In regards to regime type, debt results in a democratic advantage, as leaders can avoid the costs of war that create "republican caution."[18] Shea finds that credit allows democracies to raise more resources than they would be otherwise able to as debt, unlike other forms of war finance, avoids macroeconomic instability, and, therefore, potential disgruntled constituents.[19]

The overwhelming emphasis on debt and regime type obscures the variation in how wars can and have been paid for. States have a variety of means to raise revenue at their disposal.

They could extract resources from society directly, such as forced labor, forced savings plans, and taxation, or indirectly via debt. The state could also turn abroad for funding, procuring a sovereign-to-sovereign loan or grant, foreign debt, or plunder. The war finance strategies of the primary belligerents in the European theater during World War II highlight this variation. The US financed half its war effort through a combination of direct and indirect taxes, as well as bond campaigns. Britain engaged in taxation and bond campaigns as well, but the US–Anglo loan was integral to supplying the British and its empire. The Soviet Union engaged in debt, printing, and a loan from the US. Germany financed its war effort via forced savings plans, taxation, debt, and plunder.[20]

In addition to obscuring war finance variation, there is a tension between those studies that argue democracies devote more resources

[16] Rasler and Thompson (1983); Shea (2014); Allen and DiGiuseppe (2013).

[17] Recent studies have called Schultz and Weingast's (2003) democratic credit advantage into question. Poast (2015) finds that prior to 1914, the presence of a central bank was more influential to the government's ability to borrow during war than regime type. DiGiuseppe and Shea (2015) find nondemocratic regimes receive a higher marginal political benefit from credit compared to democratic regimes.

[18] Doyle (1986, 1160); Flores-Macías and Kreps (forthcoming).

[19] Shea (2014). See also DiGiuseppe (2015a, 2015b).

[20] For US war finance, see Murphy (1950) and Studenski and Krooss (1952). For British war finance, see Hancock and Gowing (1949), Murphy (1950), and Sayers (1956). For Soviet war finance, see Harrison (1998) and Millar (1980), and for German war finance, see Tooze (2007).

to the war effort[21] and those that suggest democracies do not.[22] Moreover, the literature that argues democracies devote more resources to the war effort finds itself in a democratic paradox. That is, democracies devote more money toward the war effort to shield citizens from human casualties,[23] yet in doing so they have to increase financial sacrifice, exposing democratic leaders to increased monitoring from a larger part of the population.

This work goes beyond the amount of resources raised and regime type to address the full variation of how wars are financed and how those means affect public support for war, security trade-offs beyond the war effort, and war fighting.

Financing the War Effort

War finance is the means by which the state meets the costs of executing the war effort. It is the manner by which the state redirects or procures monetary resources to meet government outlays to continue the war. As noted in the previous section, wars can be financed by direct or indirect resource extraction from the state's citizens or from resources outside the state.[24] To simplify the discussion I emphasize three primary forms of war finance: taxation, debt, and sovereign-to-sovereign loans.

Direct resource extraction refers to resources paid to the state directly from citizens within society. This form of extraction is imposed upon individuals or groups. That is to say, one cannot opt out of the extraction policy or transfer the burden or monetary costs to others. Taxation, particularly income or corporate taxes, is the primary form of direct resource extraction. In contrast to direct resource extraction, when the government engages in indirect resource extraction it initially asks nothing from citizens or provides citizens a choice of either opting out of the policy or transferring the costs. The primary form of indirect extraction is debt. Finally, the state can also go outside its borders to fund its war effort. External resource extraction refers to resources procured from abroad that can be used to achieve domestic

21 Fordham and Walker (2005); Goldsmith (2003, 2007).
22 Reiter and Stam (2002).
23 Gartzke (2001); Valentino et al. (2010); Caverley (2014).
24 Cappella Zielinski (2016, 20–25).

objectives.[25] While there are various forms of external extraction, I focus on sovereign-to-sovereign (S-to-S) loans from a creditor state to the belligerent in need.[26] S-to-S loans are the direct extension of funds from State A to State B.[27] Each form of research extraction has different implications for citizen interaction with and, consequently, awareness of the war effort. The extent to which citizens' sacrifice for the war effort encourages those making the sacrifice to monitor the war.[28] Thus, direct resource extraction results in a high level of wartime monitoring. In contrast, means of indirect extraction detach citizens from the feeling of sacrifice, as costs are circumvented, transferred to others, or paid in the future. As a result, the extent to which citizens are sacrificing for the war is obscured.[29] As costs are obscured, fewer citizens feel an incentive to invest in information collection regarding the

[25] Mastanduno et al. (1989); Barnett (1992). External sources of war finance have been increasingly prevalent in interstate wars since 1950. Prior to 1950, 52 percent of belligerents, for interstate wars over six months long, engaged in foreign debt, whereas 72 percent have in the postwar era. If one takes into account all forms of external war finance, the figures are more dramatic. Prior to 1950, 25 percent of states used resources from abroad to cover 25 percent or more of the cost of war. After 1950, 80 percent of states relied on resources from abroad to cover 25 percent or more of the war (Cappella Zielinski 2016, 258).

[26] I emphasize sovereign-to-sovereign grants and loans over plunder and foreign debt. Plunder is rare in the post–World War II era (see Cappella Zielinski 2016, Chapter 6). Moreover, both plunder and public debt floated on international markets do not invite geopolitical influence. Plunder does not make the state beholden to any one actor as goods are taken. In regard to financial creditors, their primary concern is profit and repayment, not geopolitical motives (Cappella Zielinski and Poast 2015, 6–9). Moreover, it is difficult for dispersed international creditors to engage in collective action (Schultz and Weingast 2003, 10–11).

[27] Most S-to-S loans are in the form of matériel, i.e., equipment and other war inputs (Cappella Zielinski and Poast 2015, 8).

[28] Lake (1992, 26); Colaresi (2014, 207). For a discussion of public ability to monitor elected officials, political responsiveness, and citizen sanctioning of leaders, see Reiter (2012). Various studies suggest that democracies, whose leaders are beholden to a larger winning coalition, are more sensitive to civilian wartime monitoring than nondemocratic regimes (Meron 2003, 21–22; Kober 2008, 2015).

[29] For example, citizens are less aware of indirect taxes, as the tax is nominally included in the price of the good or service. Thus, no explicit recognition of the payment for a public good or service enters into the individual's price adjustment. Hence the individual is likely to be ignorant of the amount of tax paid and, in some cases, unaware of the tax altogether (Cappella Zielinski 2016, 23–24). See also Buchanan (1999).

war effort.[30] At the end of this continuum are sovereign-to-sovereign loans. Whereas indirect resource extraction in the form of debt may invite some monitoring, as citizens eventually have to pay the debt, sovereign-to-sovereign loans invites even less monitoring as citizens outsource the costs to others.[31] While S-to-S loans obfuscate the costs of war from the citizenry of the belligerent state, they invite another monitoring and potential influence from the creditor state.

Public Support for War, the Military Budget, and Security beyond the War Effort

The form of resource extraction can increase public opposition to the war.[32] Taxation is politically tenuous regardless of the war effort. Individuals do not want their hard earned income confiscated. Kriner et al. find an economic self-interest mechanism suggesting that citizens do not simply weigh the benefits against the overall costs of military action. Rather, such calculations are sharply influenced by the extent to which individual citizens believe they themselves will bear the costs directly.[33] Flores-Macías and Kreps find citizens are more sensitive to the costs of war when they come into contact with the burdens of war through taxes rather than debt.[34]

However, when public support is high, this economic self-interest mechanism can be overcome. If the war is going well and citizens perceive a positive return on their financial sacrifice, then financial sacrifice is acceptable.[35] Moreover, leaders will have more room for maneuver

[30] Lake (1992, 26). See also Colaresi (2014, 206).

[31] Indirect means such as debt invites monitoring and citizen sacrifice in the long term. As Narizny (2003, 204) points out, "debt must be repaid at some point and interest will add significantly to the bill. If voters have sufficiently long time horizons, loans will still produce a serious distributional controversy . . . large-scale deficit spending can overheat an economy and cause inflation."

[32] This argument is counter to Berinsky's findings that human and monetary costs do not affect public support for war (Berinsky 2009, 82–83).

[33] Kriner et al. (2015). Kriner et al. also find that economic self-interest alone does not explain individual variation in reactions to war taxation. They find that Americans assess war taxation though a partisan lens.

[34] Flores-Macías and Kreps (forthcoming).

[35] Gelpi et al. (2009). World War II engendered consistently high levels of public support throughout nearly four years of US involvement (Berinsky 2009, 57). At the same time, World War II was financed by 50 percent taxation (Studenski

vis-à-vis policies favoring the war effort, as legislative approval is more likely in the face of strong public support[36] and presidential popularity.[37]

When public support is low, however, this self-interest mechanism comes to the fore, and individuals begin to evaluate the war effort with greater scrutiny.[38] Monitoring the war effort becomes significant as information is disseminated regarding battlefield success or failure and perceived war outcome. Those sacrificing their income expect military policy to bear the fruit of their costs.[39] Given that citizens cannot opt out of the extraction policy or transfer the labor burden or monetary costs of said policy to others, they are highly aware of the war effort. As a result, direct resource extraction increases public opposition to the war effort.[40]

Yet, even in light of low public support, wars do not necessarily end immediately.[41] Leaders face high costs, ranging from loss of office

and Krooss 1952) and the highest direct tax rates that the US has ever experienced (Internal Revenue Service 1993, 252).

36 Millett et al. (1998, 4–6); Mack (1975, 179–180); Hildebrandt et al. (2013). Tangentially, Goldsmith (2007, 197–198) argues that "taxpayers are more likely to believe that a constrained and securitized executive, operating in a more transparent institutional setting, will only levy greater taxes for legitimate reasons," which results in higher credibility, legitimacy, and trust in the state, increasing the capacity for the defense effort.

37 Edwards (1997); Krosnick and Kinder (1990); Kriner (2010, 243).

38 This argument is consistent with the casualty literature that finds that sacrifice though active service is correlated with political participation vis-à-vis the war effort (e.g., Levy 2013; Silver 2004). Often sacrifice has been discussed in terms of casualties (e.g., Gelpi et al. 2009).

39 It is worth noting, in an era of the all-volunteer force, the sacrifice of citizen lives often only reflects a small proportion of society. Given the small percentage of citizens servicing, the only connection citizens may feel toward the war is financial sacrifice. In the US, for example, according to the US Census, the active duty military population in 2010 was about 1.4 million, roughly 1 percent of the population eighteen and older (www.census.gov/prod/2011pubs/12statab/defense.pdf). It is no surprise in September 2011 "that 50 percent of (Pew) poll respondents thought that the conflicts in Iraq and Afghanistan had 'made very little difference in their life'" (Caverley 2014, 4). These finding are also supported by Silver (2004) and Vasquez (2005), who argue that as fewer people participate in the military, governments face less political opposition to preparing for, initiating, and waging war.

40 This logic reflects Gelpi et al. (2009). They find casualties do not necessarily produce a collapse in public support. Instead, casualties are filtered though expectations of success.

41 Reiter (2009); Stanley (2009); Croco (2015).

to exile or death, when they lose a war.[42] Faced with declining support, rather than end the war, elected leaders sometimes gamble for resurrection.[43] Croco finds that culpable leaders, even when knowing victory is improbable, do not want to end the conflict and prioritize their own interests at the expensive of their citizens when the state is at war. As a result, a culpable leader will have strong incentives to be an "'unfaithful agent' on behalf of the citizens if the war is not going well."[44] One recent example of gambling for resurrection is the 2007 Surge in Iraq. The decision to put more troops in the conflict came at a point when roughly 60 percent of Americans thought the war was a mistake.[45]

When exit from conflict is not an immediate option, citizens exercise their voice though the military budget. Citizens of the state react to the direct resource extraction by making calls for placing an upper bound on military spending, constraining a state's war-fighting ability[46] or decreasing resources spent in other areas in which military funds are being applied.[47] In other words, instead of the classic guns-versus-butter situation arising as the state may already be engaged in austerity measures to fund the war effort, a guns-versus-guns situation arises as an upper bound on military spending is put in place.[48]

This upper bound can be direct or indirect. Directly, representative institutions, or those groups within a society with the power of the purse, refuse to increase a state's budget authority, which allows the state to purchase new inputs for the war effort. Indirectly, the state could set limits on troop numbers, refuse to increase taxes, or

[42] See Bueno de Mesquita and Siverson (1995); Bueno de Mesquita et al. (2003); Chiozza and Goemans (2003); Colaresi (2004); Goemans (2000).

[43] Downs and Rocke (1994). [44] Croco (2015, 11). [45] Croco (2015, 1).

[46] Colaresi (2014, 206).

[47] Hindering the budget is consistent with Kriner (2010), who argues that the US Congress has few direct mechanisms to affect the war effort. Instead, it exercises its influence though indirect means, such as the introduction of legislation authorizing or seeking to curtail the use of force, oversight hearings, and engaging the debate over military policy making in the public sphere (Kriner 2010, 11–12).

[48] I build upon the guns-versus-butter framework proposed by Powell (1990). During wartime, direct extraction via taxation has a theoretical upper bound of 100 percent, although through the mechanisms discussed in this chapter, states often keep taxes below that level (see Narizny 2007, 19–20).

sanction deficit spending to pay for the war effort.[49] To cope with this environment of constrained resources, the state engages in reappropriation within the military budget. The state must make tough choices of how to spend its limited funds. Fund reappropriation can take two forms. The state can use funds appropriated for other national security goals or conflicts and redirect them to the war in question, or the state can forgo reinvestment in lost soldiers and matériel. The consequences of said reappropriation result in decreased ability for the state to confront other security threats during wartime or in the future.

For example, during the Vietnam War, to get the tax increase passed, President Johnson had to decrease overall spending. McNamara instructed the services to comb through their respective FY1968 budgets for expenditure savings. The result was deferring modernization projects and perpetuating maintenance backlogs of equipment. Funding non-Vietnam-related military programs at the previous year's level left reduced inventories unreplenished, as units steadily diminished their stockpiles to continue daily operations.[50] In early October, McNamara informed Johnson that the Department of Defense (DoD) FY1968 financial needs had increased by $4.3 billion because of the war but "internal reprogramming could cover about three-fourths of it, and that he would be able to find yet another $1 billion in 'savings' in order to submit a revised budget with zero NOA [new obligational authority] change."[51] The Revenue and Expenditure Control Act of 1968, which increased taxes, required a reduction of $10 billion in NOA and $6 billion in expenditures. Even though Vietnam operations would remain fully funded, DoD would have to reduce expenditures during FY1969 by $3 billion, further slashing non-Vietnam programs.[52] This situation worsened during the crafting of the FY1969 budget, as the services had to eliminate or defer "everything that does not contribute to [the] war" in Vietnam.[53] In addition to further cuts in big-ticket items, particularly Anti-Ballistic Missile Systems and Minuteman III programs, the Bureau of the Budget recommended drawing down seventy-five days of the supply pipeline for the Army forces in Europe.[54] During the FY1969 hearings, Congress pressed for more decreases in military expenditures. The Joint Chiefs continuously objected to decreases in spending, arguing it diminished

[49] Narizny (2003). [50] Drea (2011, 236). [51] Drea (2011, 238).
[52] Drea (2011, 251–255). [53] Drea (2011, 241). [54] Drea (2011, 243–247).

US ability to fight a European war as well as eroding general purpose forces, diminishing training and rotation base, and creating shortages in certain types of ammunition, spare parts, and aircraft, further reducing overall capabilities.

In sum, when a war becomes unpopular and it is financed by taxation, citizens express their dissatisfaction by pressuring lawmakers to constrain the military budget. To continue fighting the war the state reappropriates funds within the military budget that in turn decreases funding for other security needs beyond the war effort. Simply put, a guns-versus-guns trade-off is created.

Sovereign-to-Sovereign Loans and State Autonomy

While taxation beholdens the state to its citizens, sovereign-to-sovereign loans beholden the state to its creditor. When a state seeks inputs for war from another state, it becomes dependent on the creditor state to continue the war effort.[55] The extent of dependence varies with the need of the creditor state and on the character and amount of the war inputs received.[56] While all war inputs may be procured from abroad, military equipment, primarily finished goods such as airplanes or battleships, is the most expensive and difficult to obtain. In an era of capital-intensive, complex weapons systems, finished goods are particularly costly and supply is limited to countries possessing the most highly developed technology.[57] This dependence is compounded by the desire of leaders of the state to perform well during wartime. As noted above, there is a plethora of literature suggesting leadership survival is contingent upon successful prosecution and favorable resolution of the war effort.

This newfound dependence invites creditor states to influence debtor belligerent states by threatening to withhold future money or matériel.[58] During World War II, for example, the US took advantage of its coercive power as a creditor to achieve goals that it was unable

[55] Snyder (1984, 472). [56] Knorr (1977, 103).

[57] Knorr (1975, 50–53). Consistent with this observation, Cappella Zielinski and Poast (2015, 15–16) find that creditor states are most likely to be major powers.

[58] Drezner (2003). I do not address whether the creditor state will seek to exploit this newfound dependency nor if the creditor state is a co-belligerent in conflict. It should be expected, however, that creditor states engage in a cost–benefit analysis. They want to obtain concessions and increase their

to achieve during the interwar years.[59] As historian Warren Kimball notes,

> Although American leaders grasped the potential leverage created by British dependence upon the United States, they did not consistently use that leverage between 1937 and March 1941 . . . Once the Lend-Lease Act became law, the temptation to use it as a means of obtaining economic concessions from Britain – concessions which would promote economic multilateralism as well as directly and indirectly benefit American economic expansion – became too great to resist.[60]

Recent research suggests this creditor advantage can be used in wartime by major powers to influence the final peace settlement.[61] During World War I, for example, President Wilson used US wartime loan extension to exercise leverage over the European powers in crafting the peace after the war: "England and France have not the same views with regard to peace that we have by any means. When the war is over we can force them to our way of thinking, because by that time they will, among other things, be financially in our hands."[62]

This creditor advantage can be directed toward the war effort, as creditor states can influence belligerent policy choices. While belligerent policy choices may be influenced in a myriad of ways, one possibility is early war termination. Many have argued that anarchy in the international system results in war-termination difficulties. Even if two combatants sign an agreement to end their war, the commitment of each not to attack the other after war's end is not enforceable.[63] Third-party intervention, especially those wars in which the belligerent's continuation of the war is dependent on the third party, can provide a coercive hand to force or enforce a settlement.[64] The creditor state can coerce the belligerent state to accept an agreement it might not

power when the costs of coercion are low. For a discussion of the opportunity costs of a creditor state to influence a debtor state's behavior, see Mearsheimer (2001, 37), Nagel (1968), and Baldwin (1989, 82–99).

59 Creditor states have used their newfound dependence to extract concessions beyond the context of the war. During World War I and II, the US demanded dollar and gold reserves from the British (Horn 2002; Hall 1995, 273). During World War I, Britain engaged in similar behavior, demanding gold from France and Russia (Horn 2002, 91).

60 Kimball (1971, 241).

61 Cappella Zielinski and Poast (2015).

62 Quoted in Burk (1979, 243).

63 Reiter (2009, 3); Fearon (1995).

64 Between 1940 and 1990, 55 percent of interstate wars were resolved at the bargaining table (Walter 1997, 335).

otherwise desire by threatening to withhold funds. In essence, the creditor state forces the belligerent to reduce its war aims, which in turn affects the belligerent's strategy choices. In sum, the extension of a sovereign-to-sovereign wartime loan creates intense newfound dependences that the creditor state can exploit.

Case Study I: Truman Financing of the Korean War Effort

The Korean War was paid for by taxation.[65] Between war onset to mid 1951, high public support for the war overcame any aversion to financial sacrifice via the economic self-interest mechanism. Once casualties began to mount and a stalemate ensued, public support for the war declined and stayed consistently low for the rest of Truman's presidency.[66] After mid 1951, war taxation coupled with low public support for war, hindered the ability of the Truman administration to convince Congress to provide the necessary budget appropriations for the war effort. As a result, an upper bound on military spending was put in place that was incompatible with the needs of the war effort. Consequently, the services had to reappropriate funds from NSC-68 goals in favor of funding the war effort.[67]

The American public was initially in favor of the war and taxation as a war finance policy. Public opinion polls conducted at various points early in the war suggest strong support for even a large-scale, costly conflict.[68] An Elmo Roper survey taken just after the outbreak of hostilities revealed that 75 percent of the people supported Truman's decision to commit US forces to Korea.[69] In November 1950, a Gallup poll survey indicated that 66 percent of Americans believed the country should stay and defend South Korea, even if it meant world war.[70]

Simultaneously, survey data, telegrams to the White House, and petitions to Congress indicate that the public supported a tax increase to fund the war. In a Gallup Poll Survey in July 1950, 70 percent of respondents stated that they would be "willing to pay more money in

[65] Cappella Zielinski (2016, Chapter 2). [66] Mueller (1973, 50).

[67] For a discussion of US military spending in the context of NSC-68 prior to entry into the Korean War, see Fordham (1998). It should also be noted that while spending was constrained, the rise in military spending during the early 1950s was exponential.

[68] According to a Gallup Poll in July 1950, 57 percent of people believed that the US was "now actually in World War III" (Cappella Zielinski 2016, 76).

[69] Sandler (1995, 20). [70] Cappella Zielinski (2016, 76).

taxes to support a larger army." A month later, 51 percent of participants agreed that "federal income taxes must be increased immediately to pay for the present war in Korea and to re-arm the United States." In October 1950, about two months into the war, 51 percent of people surveyed stated that they would prefer increased taxes to pay for the war versus 26 percent who preferred to borrow.[71] Gallup reported, "Rarely has the Institute in its fifteen years of measuring public opinion found such heavy majorities expressing a willingness to pay more taxes for any public purpose."[72]

Due to World War II fiscal mobilization, taxation as a war finance strategy was not difficult to implement. Prior to war onset, Congress was in the process of reducing World War II taxes.[73] Instead of implementing new taxes, Congress eliminated various tax reductions, increased individual income tax rates to the "tentative" levels adopted in 1945 and returned to an excess profits tax.[74] High public support for the war overcame any aversion to financial sacrifice via the economic self-interest mechanism.

The Chinese invasion in November 1950, large casualties from the Chinese and North Korean offensives in the spring of 1951, and an ensuing bloody stalemate drastically hurt public support for the war.[75] In February 1951, only 48 percent of Americans thought it was right for President Truman to have sent American forces to fight in Korea.[76] Campbell et al. also found that support for Korea declined: "The Korean conflict also failed to elicit the sustained period of support evidenced in World War I and, to a slightly lesser extent, World War II. Between October 1950 and January 1951, overt support for the conflict dropped from 65 percent to an average of 31 percent."[77]

The visibility of the war finance policy coupled with growing public opposition to the war resulted in constrained military budgets.

[71] Cappella Zielinski (2016, 74). [72] Casey (2005, 672).

[73] The House of Representatives had HR 8920 under consideration, a bill that would reduce certain excise taxes, close some long-standing tax loopholes, and increase corporation income tax rates (Keith 1951, 193).

[74] Harry S. Truman to Chairman Robert Doughton, July 25, 1950, Public Papers of the President, Truman Library.

[75] On May 24, Omar Bradley revealed in congressional testimony that at that time there were 69,276 American battle casualties, which includes killed, wounded, and missing in action, and 72,679 nonbattle casualties (Blair 1987, 904).

[76] Cappella Zielinski (2016, 87). [77] Campbell et al. (1965, 324).

Citizens felt they had sacrificed enough for the war. In July 1950, 44 percent of people answered that Americans had not been asked to make enough sacrifices in support of the war. In March 1951, this figure dropped, only 20 percent of respondents feeling they had not been asked to sacrifice enough.[78] Citizen attitudes began to be reflected in Congress. When discussing the FY1952 budget, some members of Congress expressed less concern over military sufficiency than over taxpayer annoyance. George Mahon (D-TX) greeted the Secretary of Defense by telling him that there was "'much feeling among the people generally and among the Members of Congress that the military people do not conserve manpower and dollars as well as they should.' Mahon questioned whether the FY1952 military buildup should be slowed down or its emphasis changed."[79]

That public opinion was shifting vis-à-vis the war effort and that this shift was affecting the ability to spend and fund the defense budget was becoming noticeable to both the Truman administration and the Department of Defense. In a cabinet meeting in late March, Robert Lovett stated: "The Defense Department has 3 enemies – (1) Risk of visual aggression ... (2) Inflation ... (3) Relaxation. The reversal of Congressional attitude demonstrates this. In December, Congress was talking full war mobilization. It was in an attitude of gluttony. Then the attitude became to minimize the need for vital projects."[80] Two months prior, the Senate passed a universal service and training bill. When it went to the House, Representatives split up the bill and, by so doing, defeated the program.

Not only was there increased difficulty getting the preferred defense budget but also the commitment to pay for the war by taxation meant that there was an upper bound on the amount of resources that could be spent on the conflict. According to Condit, "members of Congress, whether Democratic or Republican, were in no mood to ignore the public will. Appropriations Committee Chairman Clarence Cannon made it very clear that he wanted to be 'in the black at the end of the fiscal year instead of in the red' and Defense was the major area for savings."[81]

[78] Cappella Zielinski (2016, 87). [79] Condit (1988, 257).

[80] Robert Lovett in a Cabinet Meeting, March 20, 1951, March 1951 folder, Connelly Papers Box 2, Truman Library.

[81] Condit (1988, 280).

The refusal to borrow meant that money must be saved. Given that domestic programs were already cut, Congress turned its eye to the defense budget.[82]

Negative public support for war compounded by the visibility of taxation meant constrained defense cuts for the FY1952–1953 budgets.[83] For example, PL 82–431, the Urgent Deficiency Appropriation Act signed on June 30, 1952, paid for ongoing operations, such as pay, maintenance and operations costs, and civilian relief in Korea, but did not replace items used in Korea, add to the strength of the armed forces, or facilitate future strength.[84] Replacements were particularly important. On August 15, the JCS recommended FY1952 increases of three divisions and seventeen antiaircraft Battalions for the Army. If the Korean War continued beyond December 1951, the Army would need to replace battle casualties with about sixty-eight thousand new men. Yet with no additional FY1952 money forthcoming, the military services had to manage their expansion within their allotted funds.[85]

The FY1953 budget followed a similar story. With a deficit on the rise and Congress remaining opposed to further tax increases, the situation meant decreases to the military services' budgets. Lovett warned the secretaries to "'wring the water' out of their current estimates or be 'drowned by the Congress in it.'"[86] With tax receipts estimated at only \$71 billion and the public debt expected to reach its ceiling of \$275 billion by July 1953, Truman asked for an increase in taxes.[87] Truman wanted "a-pay-as-we-go" policy and asked Congress to give him at least the shortfall in his early request for additional taxes. Instead, Congress cut the budget. In all, Congress cut the president's request by 10 percent, with the Army as the biggest loser.[88] The supplementals fared a similar fate. The first supplemental, PL 82–547, was cut by 23 percent, while the second supplemental, PL 83–11, did not raise funds at all, simply authorized the military services to transfer \$1.07 billion from already available appropriations.[89]

[82] Pierpaoli (1999, 132–134).

[83] Following war outbreak, Congress provided supplemental appropriations of \$38.2 billion for FY1951, which covered the costs of operations in Korea and general worldwide buildup of military forces. In FY1952, almost all the funding of operations in Korea was provided in regular appropriation, except for \$1.4 billion. For FY1953, Congress provided all funds for operations in Korea in regular defense appropriations bills (Daggett 2016, 3).

[84] Condit (1988, 260). [85] Condit (1988, 267). [86] Condit (1988, 275).

[87] Condit (1988, 280). [88] Condit (1988, 282). [89] Condit (1988, 282–283).

In addition to decreasing replacements for the Korean War, reappropriation decreased the financial ability of the US to meet NSC-68 objectives,[90] including NSC114–2 objectives and the ability to fight a general war against the Soviet Union.[91] For example, sensitive to the fiscal environment for FY1953 the Army and Marine Corps were instructed to prepare for a no combat assumption. Under a no combat assumption, the military services had to absorb the combat pipeline (which provided rotation and casualty replacements for Korea) within their authorized strength.[92] As a result, readiness was reduced below approved levels and material combat readiness delayed two years.[93] As a result, the Joint Chiefs qualified the military capability to carry out NSC114–2 missions. As Condit reports,

> The Army thought it could support combat operations during the first several months of general war only ... if that war continued after general war started, the total number of troops in action would drop initially because of new pipeline demands. Inability to perform all Army missions derived from insufficient reserve stocks, an inadequate mobilization, and low strength and readiness in the General Reserve. The Navy and Marine Corps rated their ability to initiate D-Day tasks as good and improving, but their ability to sustain and expand initial plans would remain unsatisfactory until they received funds to stockpile sufficient war reserves. The Air Force felt that it possessed a very limited capability to defense the Western Hemisphere and only a limited capability to carry out a strategic offensive ... Its capabilities would improve under a continued buildup.[94]

The Truman administration's financing of the Korean War via taxation faced no resistance in the early months of the war as strong public support for the war overcame any aversion to financial sacrifice. Once public support for the war declined, citizens felt that they sacrificed enough and placed an upper bound on the amount of resources that could be mobilized for the war effort. In order to cope with constrained funds, the military reappropriated funds dedicated to meeting security goals beyond the war effort.

[90] This is not to say that funding for the military services was abundant, as it was. This is to say that it was not as abundant as it otherwise would have been. For a discussion of other factors beyond reappropriation that hindered the pursuit of NSC-68 during the Korean War, see Pierpaoli (1999, Chapter 4).

[91] For discussion of the formulation of NSC114-1 and NSC114-2, see Mitrovich (2000).

[92] Condit (1988, 267). [93] Condit (1988, 289). [94] Condit (1988, 291).

Case Study II: US Material Loan to Israel during the 1973 Arab–Israeli War

The Arab–Israeli 1973 War was a financially costly war for Israel.[95] The three-week war, October 6–25, cost the equivalent of one year's GNP.[96] The state initially met these costs though a combination of taxation, debt, and austerity measures.[97] However, it was an American military loan to Israel that was crucial to sustaining the Israeli war effort.[98] From October 13 to November 12, 1973, the US provided Israel with 63,500 tons of arms and military equipment and 34 F-4s and 54 A-4s to resupply lost Israeli Phantom and Skyhawks.[99] The US also provided 2,000 antitank missiles, 14,800 rounds of 105mm APDS (armor piercing ammunition), 16,256 LAWs (light antitank weapons), 65 M-60 tanks, 23 M-109 self-propelled howitzers, and 400 M-54 trucks.[100]

Israel was reliant on the US for its military hardware in order to continue the war. There were no supplier substitutes for such advanced weaponry in large amounts that could be delivered at a moment's notice. The result of this enormous extension of war material was Israeli dependence on the US. Israel was aware that it was indebted to the desires of American policy makers as a result of the material loan. On October 12, Israel's ambassador to the US, Simcha Dinitz, told Kissinger, "Our decision whether to start a new offensive [against Egypt] or not depends on our [military] power. Without the transfer of armor, aircraft, and artillery from the United States, Israel was unable to launch a military offensive."[101]

This dependence was well recognized by American policy makers. In a Washington Special Action Group (WSAG) meeting, Secretary of

95 All documents used in this case study can be found at the National Security Archive, http://nsarchive.gwu.edu/NSAEBB/NSAEBB98/.
96 Barnett (1992, 193). The cost of the war has also been documented by Plessner (1994) and Lanir (1984).
97 Barnett (1992, 193–194). See also Gabbay (1990, 86).
98 For a discussion of US loans and grants to Israel, see Barnett (1992, 231); Rivlin (1992); Sharp (2013, 27).
99 Levey (2008, 494–495).
100 Dinitz (1999, 117). Dinitz notes that while it is difficult to estimate the financial value of the equipment supplied to Israel, President Nixon authorized a special appropriation of $2.2 billion, which was supposed to cover the cost of this equipment (Dinitz 1999, 117).
101 Quoted in Lebow and Stein (1994, 204).

State Kissinger touted Israeli dependence.[102] He stated, "We are now in a war of attrition. Without our airlift, Israel would be dead now." Later in the meeting, he went on to comment that the US could exploit this dependence to force a cease-fire: "We thought Israel preponderant, that we had to hold things back from them. Any war now is one of attrition. *Our veto is in supply*, not what they have when it starts. To get a cease-fire, we have to become engaged."[103]

A cease-fire was the primary American goal.[104] American policy makers wanted to ensure decreased Soviet interest in the region, which a prolonged war would only serve to increase, and to ensure a Soviet-supported operation would not succeed against an American-supported operation.[105] Moreover, the US did not want the Egyptian Army to be destroyed or Sadat humiliated, so that he could be party to the negotiations that would follow the war under American guidance.[106]

Knowing that Israel would not initially support a cease-fire, the US needed to foster Israeli confidence in American intentions and coerce them to accept an agreement.[107] Israeli dependency on American military supply was critical to meeting these goals. On October 17, President Nixon opened the meeting with the statement, "In order to have the influence we need to bring Israel to a settlement, we have to have their confidence." He went on to say, "The second point is that we have to do enough to have a bargaining position to bring Israel kicking and screaming to the table."[108]

[102] Members of the WSAG included President Nixon, Secretary of State Henry Kissinger, Deputy Secretary of State Ambassador Kenneth Rush, Deputy Secretary of Defense William Clements, Director of the CIA William Colby, Chairman of the Joint Chiefs of Staff Admiral Thomas Morrer, Assistant Secretary of State for Near Eastern and South Asian Affairs Joseph Sisco, and Deputy Assistant to the President for National Security Affairs Major General Brent Scowcroft.

[103] Minutes, "Washington Special Action Group Meeting," October 17, 1973, 3:05 P.M.–4:04 P.M., NPMP, NSC Institutional Files, box H-117, WSAG Minutes (originals), 10–2–73 to 7–23–74 (2 of 3), emphasis added.

[104] For a discussion of the importance of the timing and conditions of the cease-fire to Kissinger, see Lebow and Stein (1994, 205).

[105] For a discussion of various American policy goals during the 1973 war, see Kissinger (1979, 564).

[106] Dinitz (1999, 113–114).

[107] For a discussion of various cease-fire attempts, see Lebow and Stein (1994, Chapter 9).

[108] See Nixon opening statements at the October 17 WSAG meeting. Memcon, "WSAG Principles: Middle East War," October 17, 1973, 4:00 P.M., NPMP,

By mid October, however, it looked as if a cease-fire would not be feasible as Israel had the opportunity to demolish Egypt's Third Army. According to a State Department Situation Report on October 19, the Israeli Defense Force stated,

> Israeli military planners may well decide that by pushing toward Cairo, this will force Egypt to withdraw units form the integrated defensive system it maintains along the Canal. This would make Israel's declared objective of destroying Egypt's army more easily obtainable. Israelis feel they have now turned the corner in the war and that the initiative on both fronts is now in Israel's hands. The smell of victory will not make Israel welcome a ceasefire.[109]

American policy makers, however, were eager to prevent Israel's pursuit of the Egyptian Army and continue with the cease-fire. On October 20, Scowcroft advised Kissinger, "The current Israeli success at Suez must not deflect us from going all out to achieve a just settlement now."[110] Kissinger responded in kind, "Our first objective must be a ceasefire. That will be tough enough to get the Israelis to accept."[111]

A cease-fire agreement between Kissinger and Moscow was reached and approved by the United Nations Security Council on October 22. However, Israel took advantage of continued firing to encircle Suez and seal the closing on the Egyptian Third Army.[112] American policy makers feared that the destruction of the Third Army would bring collapse of Sadat's regime.[113] Moreover, they feared that it would bring about a confrontation with the Soviets.

Given that fighting was continuing despite cease-fire efforts, the US begin to increase pressure on Israel.[114] Kissinger asked Alexander Haig

NSC Institutional Files, box H–92, WSAG Meeting Middle East 10/17/73, folder 6.

[109] Department of State, Operations Center, Middle East Task Force, Situation Report 43, "Situation Report in the Middle East as of 0600 Hours EDT, Oct. 19, 1973," NPMP, NSCF, box 1173, 1973 Middle East War, October 19, 1973, File No. 14.

[110] Situation Room Message from Peter Rodman to Kissinger, TOHAK 20, October 20, 1973, transmitting memorandum from Scowcroft to Kissinger, NPMP, HAKO, box 39, HAK Trip – Moscow, Tel Aviv, London, October 20–23, 1973, TOHAK 1–60.

[111] Message from Kissinger to Scowcroft, HAKTO 06 [October 20, 1973], NPMP, HAKO, box 39, HAK Trip – Moscow, Tel Aviv, London, October 20–23, 1973, HAKTO, SECTO, TOSEC, Misc.

[112] For a discussion of the failure of Kissinger to send a clear message to the Israelis regarding the cease-fire, see Lebow and Stein (1994, 218).

[113] Dinitz (1999, 119). [114] Lebow and Stein (1994, 230–231).

to call Ambassador Dinitz and demand, on behalf of the president, an immediate end to offensive military action by Israel.[115] On October 24, Nixon wrote to Brezhnev "that the Israeli Government was informed that any further offensive operations would lead to a severe deterioration of relations between the Israeli and US government" and that "he personally intervened with the PM of Israel to halt the fighting. Specifically, at no place since the beginning of the ceasefire at 7:00am today have Israelis tried to advance."[116]

In a briefing memo from Assistant Secretary of State for Intelligence and Research Ray Cline to Kissinger on October 24, Cline noted the differing motivations for accepting the cease-fire. "The ceasefire intervened just as Israel appeared to be well on the way to fragmenting the Egyptian forces on the west bank and isolating those on the east bank. Apparently, Israel halted its victory drive only out of deference to Washington and has no real interest in letting Cairo get off 'so lightly.'"[117] Kissinger continued his message of American influence to Egypt. Kissinger met with Acting Egyptian Foreign Minister Ismail Fahmi. He told Fahmi, concerned about the potential for Israeli military advancement to Cairo, that Israel would submit American influence. "They will listen to us. Israel needs us. It needs American supplies."[118] The cease-fire commenced on October 25.

Relying on the US for a military loan resulted in Israeli dependence for American goods. American policy makers used this dependence to shape the peace process. Specifically, the US influenced the Israelis to curb their offensive strategy and preserve the Egyptian Third Army in order to pursue a cease-fire.

Conclusion

How a state chooses to finance its war has repercussions beyond the financial costs of war. This chapter identified various aspects of war

[115] Lebow and Stein (1994, 231).

[116] Scowcroft letter to Dobrynin, enclosing message from Nixon to Brezhnev, October 24, 1973, delivered to Soviet Embassy, 1:00 P.M., NPMP, HAKO, box 69, Dobrynin/Kissinger Vol. 20 (October 12–November 27, 1973).

[117] Ray Cline, Assistant Secretary of State for Intelligence and Research, to Kissinger, "Ceasefire Problems," October 24, 1973, RG 59, SN 70-73 POL 27-14 Arab-Isr.

[118] Memcon between Kissinger and Fahmi, October 30, 3:08 P.M., RG 59, SN 70-73, POL 27 Arab-Isr.

finance that can, on the margin, affect the ability of the state to pursue its war effort. The manner of war finance can result in costly trade-offs regarding public support for the war, confronting security needs beyond the war effort, and war fighting. When public support for war becomes negative and the war is financed by taxation a guns-versus-guns trade-off arises. Given that citizens cannot sue for peace, they indirectly constrain the state by limiting the military budget. To continue fighting the war, the state reappropriates funds within the military budget. As a result of this guns-versus-guns trade-off, the state decreases funding for other security needs in order to meet the needs of the war effort. In contrast, external extraction results in belligerent dependence on a creditor state. This dependence allows the creditor state to influence the belligerent, including early war termination.

The findings of this chapter imply that, ceteris paribus, states able to obtain resources for war from diffuse sources will be able to sustain larger and longer wars without facing costly trade-offs. More succinctly put, borrowing is the most effective method of war finance. Given its importance, it would behoove scholars to explore under what conditions states can float debt successfully. Specifically, the domestic institutions, i.e., central banks and state treasuries, that manage war debt. Moreover, given new war finance data sets and expanding data on sovereign debt yields and financial institutions, future researchers should engage in more systemic tests of the various forms of war finance on military effectiveness.

References

Allen, Michael A., and Matthew DiGiuseppe. 2013. Tightening the Belt: Sovereign Debt and Alliance Formation. *International Studies Quarterly* 57(4): 647–659.

Arreguín-Toft, Ivan. 2005. *How the Weak Win Wars: A Theory of Asymmetric Conflict*. Cambridge: Cambridge University Press.

Baldwin, David A. 1989. *Paradoxes of Power*. New York: Basil Blackwell.

Barnett, Michael N. 1992. *Confronting the Costs of War: Military Power, State, and Society in Egypt and Israel*. Princeton, NJ: Princeton University Press.

Beckley, Michael. 2010. Economic Development and Military Effectiveness. *Journal of Strategic Studies* 33(1):43–79.

Berinsky, Adam. 2009. *In Time of War: Understanding American Public Opinion from World War II to Iraq*. Chicago: University of Chicago Press.

Betts, Richard K. 1995. *Military Readiness: Concepts, Choices, Consequences*. Washington, DC: Brookings Institution Press.

Biddle, Stephen. 2004. *Military Power: Explaining Victory and Defeat in Modern Battle*. Princeton, NJ: Princeton University Press.

Blair, Clay. 1987. *The Forgotten War: American in Korea 1950–1953*. New York: Times Books.

Brooks, Risa A. 2003. Making Military Might: Why Do States Fail and Succeed? A Review Essay. *International Security* 28(2):149–191.

2007. Introduction: The Impact of Culture, Society, Institutions, and International Forces on Military Effectiveness. In R. A. Brooks and E. A. Stanley, eds., *Creating Military Power: The Sources of Military Effectiveness*, 1–26. Stanford, CA: Stanford University Press.

Buchanan, James M. 1999 (1967). *The Collected Works of James M. Buchanan, Vol. 4. Public Finance in Democratic Process: Fiscal Institutions and Individual Choice*. Indianapolis, IN: Liberty Fund.

Bueno de Mesquita, Bruce, and Randolph M. Siverson. 1995. War and the Survival of Political Leaders: A Comparative Study of Regime Types and Political Accountability. *American Political Science Review* 89(4): 841–855.

Bueno de Mesquita, Bruce, Alastair Smith, Randolph M. Siverson, and James Morrow. 2003. *The Logic of Political Survival*. Cambridge, MA: MIT Press.

Burk, Kathleen. 1979. Great Britain in the United States, 1917–1918: The Turning Point. *The International History Review* 1(2): 228–245.

Campbell, Joel T., Leila S. Cain, Anatol Rapoport, and Phillip E. Converse. 1965. Public Opinion and the Outbreak of War. *Journal of Conflict Resolution* 9(3): 318–333.

Cappella Zielinski, Rosella. 2016. *How States Pay for Wars*. Ithaca, NY: Cornell University Press.

Cappella Zielinski, Rosella, and Paul Poast. 2015. Financial Intervention and Conflict: Explaining Sovereign-to-Sovereign Loans to War Belligerents. Working Paper, Boston University.

Carter, Jeff, and Glenn Palmer. 2015. Keeping the Schools Open While the Troops Are Away: Regime Type, Interstate War, and Government Spending. *International Studies Quarterly* 59(2): 145–157.

Casey, Steven. 2005. Selling NSC-68: The Truman Administration, Public Opinion, and the Politics of Mobilization, 1950–1951. *Diplomatic History* 29(September): 655–690.

Caverley, Jonathan D. 2014. *Democratic Militarism: Voting, Wealth and War*. Cambridge: Cambridge University Press.

Centeno, Miguel A. 1997. Blood and Debt: War and Taxation in Nineteenth-Century Latin America. *American Journal of Sociology* 102(May): 1565–1605.

Chiozza, Giacomo, and H. E. Goemans. 2003. Peace through Insecurity: Tenure and International Conflict. *Journal of Conflict Resolution* 47(4): 443–467.

Colaresi, Michael P. 2004. When Doves Cry: International Rivalry, Unreciprocated Cooperation, and Leadership Turnover. *American Journal of Political Science* 48(3): 555–570.

2014. *The Secrecy Dilemma in National Security*. Oxford: Oxford University Press.

Condit, Doris M. 1988. *The Test of War 1950–1953*. Washington, DC: Historical Office, Office of the Secretary of Defense.

Croco, Sarah E. 2015. *Peace at What Price? Leader Culpability and the Domestic Politics of War Termination*. Cambridge: Cambridge University Press.

Daggett, Stephen. 2016. *CRS Report for Congress, Military Operations: Precedents for Funding Contingency Operations in Regular or in Supplemental Appropriations Bills*. Washington, DC: Congressional Research Service, Library of Congress.

DiGuiseppe, Matthew. 2015a. The Fiscal Autonomy of Deciders: Creditworthiness and Conflict Initiation. *Foreign Policy Analysis* 11(3): 317–338.

2015b. Guns, Butter, and Debt: Sovereign Creditworthiness and Military Spending. *Journal of Peace Research* 52(2): 680–693.

DiGiuseppe, Matthew, and Patrick E. Shea. 2015. Sovereign Credit and the Fate of Leaders: Reassessing the "Democratic Advantage." *International Studies Quarterly* 59(September): 447–570.

Dinitz, Simcha. 1999. The Yom Kippur War: Diplomacy of War and Peace. *Israel Affairs* 6(1): 104–126.

Downs, George W., and David M. Rocke. 1994. Conflict, Agency, and Gambling for Resurrection: The Principal-Agent Problem Goes to War. *American Journal of Political Science* 39(May): 362–380.

Doyle, Michael. 1986. Liberalism and World Politics. *American Political Science Review* 80(4): 1151–1169.

Drea, Edward J. 2011. *McNamara, Clifford, and the Burdens of Vietnam 1965–1969*. Washington, DC: Historical Office of the Secretary of Defense.

Drezner, Daniel. 2003. The Hidden Hand of Economic Coercion. *International Organization* 57(3): 643–659.

Edwards, George. 1997. Aligning Tests with Theory: Presidential Approval as a Source of Influence in Congress. *Congress and the Presidency* 24(2): 113–130.

Fearon, James D. 1995. Rationalist Explanations for War. *International Organization* 49(Summer): 379–414.

1999. Electoral Accountability and the Control of Politicians: Selecting Good Types versus Sanctioning Poor Performance. In Adam Przeworski, Susan C. Stokes, and Bernard Manin, eds., *Democracy, Accountability, and Representation*, 55–97. Cambridge: Cambridge University Press.

Flores-Macías, Gustavo A., and Sarah E. Kreps. 2013. Political Parties at War: A Study of American War Finance. *American Political Science Review* 107(4): 833–848.

Forthcoming. Borrowing Support for War: The Effect of War Finance on Public Attitudes toward Conflict. *Journal of Conflict Resolution.*

Fordham, Benjamin O. 1998. *Building the Cold War Consensus: The Political Economy of US National Security Policy, 1949–1941*. Ann Arbor: University of Michigan Press.

Fordham, Benjamin O., and Thomas C. Walker. 2005. Kantian Liberalism, Regime Type, and Military Resource Allocation: Do Democracies Spend Less? *International Studies Quarterly* 49(1): 141–157.

Gabbay, Yosef. 1990. Israel's Fiscal Policy, 1948–1982. In Moshe Sanbar, ed., *Economic and Social Policy in Israel: The First Generation*, 85–112. Lanham, MD: University Press of America/Jerusalem Center for Public Affairs/Center for Jewish Community Studies.

Gartzke, Erik. 2001. Democracy and the Preparation for War: Does Regime Type Affect States' Anticipation of Casualties? *International Studies Quarterly* 45(3): 467–484.

Gelpi, Christopher, Peter Feaver, and Jason Reifler. 2009. *Paying the Human Costs of War: American Public Opinion and Casualties in Military Conflicts*. Princeton, NJ: Princeton University Press.

Gerring, John. 2017. *Case Study Research: Principles and Practices*. 2nd ed. Cambridge: Cambridge University Press.

Gilpin, Robert. 1981. *War and Change in World Politics*. Cambridge: Cambridge University Press.

Goemans, H. E. 2000. *War and Punishment*. Princeton, NJ: Princeton University Press.

Goldsmith, Benjamin E. 2003. Bearing the Defense Burden, 1886–1989: Why Spend More? *Journal of Conflict Resolution* 47(5): 551–573.

2007. Defense Effort and Institutional Theories of Democratic Peace and Victory: Why Try Harder? *Security Studies* 16(2): 189–222.

Hall, H. D. 1995. *North American Supply*. London: Her Majesty's Stationery Office.

Hancock, W. K., and M. M. Gowing. 1949. *British War Economy*. London: His Majesty's Stationery Office.

Harrison, Mark. 1998. The Economics of World War II: An Overview. In Mark Harrison, ed., *The Economics of World War II*, 1–42. Cambridge: Cambridge University Press.

Harrison, Todd. 2014. Rethinking Readiness. *Strategic Studies Quarterly* 8(3): 28–68.

Harsanyi, J. C. 1962. Measurement of Social Power, Opportunity Costs, and the Theory of Two-Person Bargaining Games. *Behavioral Science* 7(1): 67–80.

Hildebrandt, Timothy, Courtney Hillebrecht, Peter M. Holm, and Jon Pevehouse. 2013. The Domestic Politics of Humanitarian Intervention: Public Opinion, Partisanship, and Ideology. *Foreign Policy Analysis* 9(3): 243–266.

Horn, Martin. 2002. *Britain, France, and Financing*. Montreal: McGill-Queen's University Press.

Howell, William G., and Jon C. Pevehouse. 2007. *While Dangers Gather: Congressional Checks on Presidential War Powers*. Princeton, NJ: Princeton University Press.

Hozier, H. M. 2005 (1878). *The Russo Turkish War*. Vol. IV. London: Elibron Classics (William Mackenzie).

Internal Revenue Service. 1993. *IRS Historical Fact Book: A Chronology 1646–1992*. Washington, DC: US Department of the Treasury.

Keith, E. G. 1951. The Excess Profits Tax Act of 1950. *National Tax Journal* 4(3): 193–207.

Kimball, Warren F. 1971. Lend-Lease and the Open Door: The Temptation of British Opulence, 1937–1942. *Political Science Quarterly* 86(2): 232–259.

Kissinger, Henry. 1979. *White House Years*. Boston: Little, Brown.

Knorr, Klaus. 1975. *The Power of Nations: The Political Economy of International Relations*. New York: Basic Books.

1977. International Economic Leverage and Its Uses. In K. Knorr and F. N. Trager, eds., *Economic Issues and National Security*, 99–126. Lawrence: Regents Press of Kansas.

Kober, Avi. 2008. The Israel Defense Forces in the Second Lebanon War: Why the Poor Performance? *The Journal of Strategic Studies* 31(1): 3–40.

2015. From Heroic to Post-Heroic Warfare: Israel's Way of War in Asymmetrical Conflicts. *Armed Forces and Society* 41(1): 96–122.

Kriner, Douglass. 2010. *After the Rubicon: Congress, Presidents, and the Politics of Waging War*. Chicago: University of Chicago Press.

Kriner, Douglass, Breanna Lechases, and Rosella Cappella Zielinski. 2015. Self-Interest, Partisanship, and the Conditional Influence of Taxation on Support for War in the United States. *Conflict Management and Peace Science*, November 24, 1–22.

Krosnick, Jon A., and Donald R. Kinder. 1990. Altering the Foundations of Support for the President through Priming. *American Political Science Review* 84(2): 497–512.

Kugler, Jacek, and William Domke. 1986. Comparing the Strength of Nations. *Comparative Political Studies* 19(1): 36–69.

Lake, David. 1992. Powerful Pacifists: Democratic States and War. *American Political Science Review* 86(1): 24–37.

Lamborn, Alan C. 1983. Power and the Politics of Extraction. *International Studies Quarterly* 27(June): 125–146.

Lanir, Zvi. 1984. Political Aims and Military Objectives – Some Observations on the Israeli Experience. In Zvi Lanir, ed., *Israeli Security Planning in the 1980s: Its Politics and Economics*, 14–49. New York: Praeger.

Lebow, Richard Ned, and Janice G. Stein. 1994. *We All Lost the Cold War*. Princeton, NJ: Princeton University Press.

Levey, Zach. 2008. Anatomy of an Airlift: United States Military Assistance to Israel during the 1973 War. *Cold War History* 8(4): 481–501.

Levi, Margaret. 1988. *Of Rule and Revenue*. Berkeley: University of California Press.

Levy, Yagil. 2013. How Military Recruitment Affects Collective Action and Its Outcomes. *International Studies Quarterly* 57(1): 28–40.

Lobell, Steven E., Norrin M. Ripsman, and Jeffery W. Taliaferro, eds. 2009. *Neoclassical Realism, the State, and Foreign Policy*. Cambridge: Cambridge University Press.

Mack, Andrew. 1975. Why Big Nations Lose Small Wars: The Politics of Asymmetric Conflict. *World Politics* 27(2): 175–200.

Mastanduno, Michael, David Lake, and John Ikenberry. 1989. Toward a Realist Theory of State Action. *International Studies Quarterly* 33(4): 457–474.

Mearsheimer, John. 2001. *The Tragedy of Great Power Policies*. New York: W. W. Norton.

Meron, Gil. 2003. *How Democracies Lose Small Wars*. Cambridge: Cambridge University Press.

Millar, James R. 1980. Financing the Soviet War Effort in World War II. *Soviet Studies* 32(1): 106–123.

Millett, Allan R., W. Murray, and Kenneth H. Watman. 1988. The Effectiveness of Military Organizations. In A. R. Millett and W. Murray, eds., *Military Effectiveness: Vol. 1. The First World War*, new ed., 1–30. Cambridge: Cambridge University Press.

Mitrovich, Gregory. 2000. *Undermining the Kremlin: America's Strategy to Subvert the Soviet Block, 1947–1956*. Ithaca, NY: Cornell University Press.

Mueller, John E. 1973. *War, Presidents and Public Opinion*. New York: John Wiley.

Murphy, H. C. 1950. *The National Debt in War and Transition*. New York: McGraw-Hill.

Nagel, Jack H. 1968. Some Questions about the Concept of Power. *Systems Research and Behavioral Science* 13(2): 129–137.

Narizny, Kevin. 2003. Both Guns and Butter, or Neither: Class Interests in the Political Economy of Rearmament. *American Political Science Review* 97(2): 203–220.

2007. *The Political Economy of Grand Strategy*. Ithaca, NY: Cornell University Press.

Organski, A. F. K., and Jacek Kugler. 1980. *The War Ledger*. Chicago: University of Chicago Press.

Pierpaoli, Paul G. 1999. *Truman and Korea: The Political Culture of the Early Cold War*. Ann Arbor: University of Michigan Press.

Plessner, Yakir. 1994. *The Political Economy of Israel from Ideology to Stagnation*. SUNY Series in Israeli Studies. Albany: State University of New York Press.

Poast, Paul. 2015. Central Banks at War. *International Organization* 69(1): 63–95.

Powell, Robert. 1990. *In the Shadow of Power: States and Strategies in International Politics*. Princeton, NJ: Princeton University Press.

Rasler, Karen A., and William R. Thompson. 1983. Global Wars, Public Debts, and the Long Cycle. *World Politics* 35(4): 489–516.

Reinhardt, Carmen M., and Christoph Trebesch. 2015. The Pitfalls of External Dependence: Greece, 1829–2015. NBER Working Paper 21664, National Bureau of Economic Research, Cambridge, MA.

Reiter, Dan. 2009. *How Wars End*. Princeton, NJ: Princeton University Press.

2012. Democracy, Deception, and Entry into War. *Security Studies* 21(4): 594–623.

Reiter, Dan, and Allan C. Stam. 2002. *Democracies at War*. Princeton, NJ: Princeton University Press.

Rivlin, Paul. 1992. *The Israeli Economy*. Boulder, CO: Westview Press.

Sandler, Stanley, ed. 1995. *The Korean War: An Encyclopedia*. New York: Routledge.

Sayers, R. S. 1956. *Financial Policy 1939–1945*. London: Her Majesty's Stationery Office.

Schultz, Kenneth A., and Barry R. Weingast. 2003. The Democratic Advantage: Institutional Foundations of Financial Power in International Competition. *International Organization* 57(Winter): 3–42.

Seawright, Jason, and John Gerring. 2008. Case Selection Techniques in Case Study Research: A Menu of Qualitative and Quantitative Options. *Political Research Quarterly* 61(2): 294–308.

Sharp, Jeremy M. 2013. *CRS Report for Congress, US Foreign Aid to Israel*. Washington, DC: Congressional Research Service, Library of Congress.

Shea, Patrick. 2014. Financing Victory: Sovereign Credit, Democracy, and War. *Journal of Conflict Resolution* 58(5): 771–795.

Silver, Beverly. 2004. Labor, War, and World Politics: Contemporary Dynamics in World-Historical Perspective. In Berthold Unfried, Marcel Van der Linden, and Christine Schindler, eds., *Labour and New Social Movements in a Globalizing World System*, 19–38. Leipzig: Akademische Verlaganstalt.

Snyder, Glenn H. 1984. The Security Dilemma in Alliance Politics. *World Politics* 36(4): 461–495.

Stam, Allan C. 1996. *Win, Lose, or Draw: Domestic Politics and the Crucible of War*. Ann Arbor: University of Michigan Press.

Stanley, Elizabeth A. 2009. *Paths to Peace: Domestic Coalition Shifts, War Termination and the Korean War*. Stanford, CA: Stanford University Press.

Studenski, Paul, and Herman E. Krooss. 1952. *Financial History of the United States*. New York: McGraw-Hill.

Tellis, Ashley J., Janice Bially, Christopher Layne, Melissa McPherson, and Jerry M. Sollinger. 2000. *Measuring National Power in the Postindustrial Age*. Santa Monica, CA: RAND Corporation.

Tooze, Adam. 2007. *The Wages of Destruction: The Making and Breaking of the Nazi Economy*. New York: Viking.

Valentino, Benjamin A., Paul K. Huth, and Sarah E. Croco. 2010. Bear Any Burden? How Democracies Minimize Costs of War. *Journal of Politics* 72(2): 528–544.

Vasquez, Joseph Paul. 2005. Shouldering the Soldiering: Democracy, Conscription, and Military Casualties. *Journal of Conflict Resolution* 49(6): 849–873.

Walter, Barbara F. 1997. The Critical Barrier to Civil War Settlement. *International Organization* 51(3): 335–364.

4 *Forced to Fight*

Coercion, Blocking Detachments, and Trade-offs in Military Effectiveness

JASON LYALL

It takes a very brave man to be a coward in the Red Army.
– Marshall Georgi Zhukov

The Islamic State's (ISIS) rapid march through northern Iraq during summer 2014 caught most observers by surprise. Its ability to recruit locally and internationally, along with its substantial combat power, prompted an American return to Iraq in the form of a sustained air campaign. Much less attention, however, has been cast on ISIS's mechanisms for maintaining discipline within its own ranks, especially after its momentum was halted around Kobane during winter 2014. ISIS, like many combatants before it, created special units dedicated to enforcing unit cohesion – in particular, by using the threat of violent sanction – to reduce the odds of desertion and defection and to tamp down factionalism arising from the ethnically mixed nature of its forces. These ISIS units have executed hundreds of soldiers caught deserting or defecting; have arrested hundreds more, often along with their families; and have positioned themselves on the battlefield specifically to catch foreign fighters seeking to escape via Turkey or to defect to rival militant organizations. As one soldier noted, "ISIS wants to kill everyone who says, 'no' [to it]. Everyone must be with them. If you turn against ISIS, they will kill you."[1]

Our existing theories of soldier motivation, as well as leading historical accounts of key wars, typically dismiss the use of coercion as ineffective, amoral, and confined to a handful of "deviant" combatants. Yet so-called blocking detachments like those used by ISIS have a long history in warfare; indeed, as detailed below, nearly one-third of all belligerents in post-1800 wars have deployed "blocking,"

[1] *International Business Times* (2014); *Financial Times* (2014); *New York Times* (2015); Reuters (2015); *Independent* (2015).

"barrier," or "antiretreat" formations. These units participated in some of the most important wars of the past two hundred years, including the Napoleonic Wars, the Taiping and Nien Rebellions in China, World War I, the Russian Civil War, and the Warlord Era in China (1916–1928). Perhaps most well known is the use of blocking detachments (*zagraditel'nye otriady*) by the Red Army during the Second World War. Over 158,000 Soviet soldiers would die at the hands of their own comrades in 1941–1944.[2]

Blocking detachments also offer a window into the trade-offs that exist across different facets of military effectiveness at the war-fighting level. While earlier conceptual frameworks do acknowledge the possibility of trade-offs across and within levels of analysis,[3] most current theorizing instead tends to focus on discrete "building blocks" of military effectiveness. As Dan Reiter acknowledges in this volume's introduction, this emphasis on narrow aspects of military effectiveness, however important these elements individually, tends to obscure tough decisions by combatants about which aspects of military effectiveness to privilege and which to sacrifice in pursuit of broader political aims. Put differently, military effectiveness is a spectrum of tasks, and proficiency in one area may come at the expense of another. At least for some belligerents, not all good things go together.

These units impose a central trade-off: coercion can artificially bolster a military's staying power – that is, its cohesion and resolve once battlefield fortunes turn against it – but at the cost of increased casualties and new vulnerabilities that can worsen its loss–exchange ratio in combat. This trade-off manifests itself across several dimensions of military effectiveness. Blocking detachments can forestall (further) desertion and defection, for example, but using violence against one's own soldiers obviously contributes to the butcher's bill of casualties. Their presence can strengthen command and control (C2) over wayward soldiers and their officers. In doing so, however, they also generate incentives to rely on rigid tactics and operations that increase vulnerability to enemy fire, again increasing losses. These detachments can bolster soldier resolve through the threat of punishment, but soldier grievances will also mount, worsening morale and possibly spilling over into greater insubordination, including officer executions ("fragging").

2 Daines (2008); Telitsyn (2010); Orlov (2012).
3 Millett and Murray (1988).

In keeping with this volume's themes, I explore the trade-offs inherent in blocking detachments at the war-fighting level in conventional wars. I first provide descriptive statistics and context on the historical use of blocking detachments using Project Mars, a new data set of 825 combatant observations spread across 250 conventional wars (1800–2011).[4] I then outline four possible trade-offs in their battlefield deployment and illustrate them using cases drawn from Project Mars. I also conduct an initial plausibility probe of these trade-offs using process tracing by drawing on Soviet experiences at Stalingrad (1942–1943) and Kursk (1943).[5] Next, I consider several additional trade-offs that emerge from this process tracing before concluding with suggestions for future research on coercion, blocking detachments, and military effectiveness.

Blocking Detachments in History

I define *blocking detachments* as specialized armed formations that are authorized to monitor, coerce, and lethally sanction an army's own officers and soldiers to ensure unit cohesion and discipline during wartime. These formations have five properties. First, while they may owe their origins to informal practices adopted haphazardly by frontline commanders, these formations are formally authorized by senior military commanders. Second, these units are typically stationed in the immediate rear of deployed forces to guard against unauthorized withdrawal and to prod soldiers into action when attacking. These units do not usually engage enemy forces, instead saving their fire for their own forces. Third, these units possess both the formal authorization and military capacity to threaten and punish soldiers in several ways. Possible sanctioning mechanisms include coerced return to units after successful desertion, forced enrollment in penal battalions, and execution, oftentimes in front of a soldier's own unit. In some cases, blocking detachments have the reach to extend punishment to a soldier's family. Fourth, blocking detachments act as barriers between the soldiers and the rear area, inhibiting the flow of information to and from soldiers while also limiting their chances of escape. Finally, while exceptions do

[4] For a detailed overview of Project Mars, see Lyall (2016a).

[5] On process tracing, see George and Bennett (2005). For process tracing in civil war settings, see Lyall (2015).

exist,[6] these units are staffed by personnel chosen for their presumed regime loyalty. Their diversion from frontline roles thus represents a costly investment since these hardline supporters are often the most effective units available to military authorities.

Despite these shared traits, the institutional design of these units has varied somewhat across combatants. For example, their coverage net can vary substantially: in some armies, only a handful of blocking detachments are deployed, often positioned behind the most unreliable units, while in other cases, they are stationed behind all frontline units. The size of blocking detachments can also vary considerably, with some units as small as fifty soldiers and others as large as two hundred or more men. Staffing, too, differs: in some cases, these units are drawn from regular military formations; in others, specialized agencies (such as military intelligence organs); and in some cases, both types of units are present, often cross-guarding each other. Finally, blocking detachments vary in the amount and frequency of violence directed against fellow soldiers. Why this heterogeneity is observed across belligerents is an important question in its own right.[7]

Most discussions of these units, if they occur at all, consist of scattered references buried within broader accounts of military effectiveness. To study blocking detachments systematically, I gathered data on the use of these units in 250 conventional wars from 1800 to 2011. This data set, known as Project Mars, encompasses 825 observations from 229 unique belligerents in conventional wars involving two or more states that resulted in ≥500 battle deaths. Belligerents are included if they possess a political capital, the ability to control their population, can muster a conventional army, and had ≥1 percent of the total fielded forces or casualties in a war. Civil wars that were fought

[6] The Zhili Clique used a special unit of child soldiers (Du Jun Dui) to shoot deserters with cannon fire in its 1925 war against the Fengtian Clique in China, for example.

[7] For coding purposes, Project Mars did not impose a numerical threshold for inclusion as a blocking detachment. Instead, three inclusion criteria were used: (1) formal orders creating the units from senior military commanders; (2) actual deployment on the battlefield; and (3) the execution of soldiers, whether in a formal setting (i.e., a tribunal) or shooting (or artillery bombardment) in the heat of battle to stem desertion, block retreat, or force soldiers forward. Here, too, we did not impose a specific number of soldier deaths, partly owing to variation in the size of militaries across wars and their associated historiographies.

conventionally – that is, with firearm-equipped uniformed soldiers engaging in direct combat using combined arms – are also included. These civil wars include the American Civil War, the Taiping and Nien Rebellions in China, the Russian Civil War, and the Spanish Civil War, among others.

What do these data reveal? Belligerents deployed blocking detachments at least 158 times, representing nearly 20 percent of all combatant observations (158/825). There were 87 recorded instances of blocking unit deployment in 483 pre–World War I observations (18 percent of all observations); an additional 71 instances were recorded in the post–World War I era (71/342, or 21 percent).[8] These data indicate that armies were equally likely to resort to blocking detachments across the "premodern" (1800–1917) and post–World War I "modern" (1918–) eras of warfare, suggesting that these units are not simply a response to the increased lethality of modern warfare.[9] Nor are these units confined to the same set of combatants. While some countries have relied heavily on such practices (notably, Russia and China), nearly 34 percent of all combatants in the data set recorded at least one deployment of blocking detachments in the 1800–1917 era (46/137). A similar 25 percent of combatants in the post-1917 era (31/124) did so as well.

Several trends emerge from these data. Above all, the deployment of blocking detachments is strongly associated with a regime's prewar repression of ethnic groups within its national borders. As I argue elsewhere,[10] regimes ruling divided societies are likely to adopt coercive approaches to generating combat power because their use of violence in the prewar era creates grievances among targeted populations that translates into suspected political disloyalty by soldiers drawn from these populations. A legacy of prewar marginalization or regime-imposed violence is also unlikely to create units with high levels of motivation. Instead, prior violence hardens ethnic group identification, fueling collective action such as desertion and defection designed

[8] These are conservative estimates and likely undercount the actual number of combatants deploying blocking units given gaps in the historical record and the tendency of some belligerents to suppress or purge these accounts from official histories.

[9] On the "modern system," see Biddle (2004). On the importance of periodization when testing our theories of war and combat, see Lyall and Wilson (2009).

[10] Lyall (2016a, 2016b).

to escape the state's military machine. Coercion thus becomes an attractive – and perhaps the only – option for generating and sustaining combat power since patriotic appeals and selective incentives are unlikely to trump soldiers' affiliation with their (repressed) ethnic group. The Ottoman Empire and Russia/Soviet Union, along with lesser-known combatants, such as the Central Asian khanate of Kokand and the Mahdist State, all resorted to blocking detachments to monitor and sanction marginalized groups with prior exposure to regime-directed repression.

While it might be supposed that blocking detachments are adopted only as last-ditch defensive measures by desperate belligerents, these data reveal that initiators (18 percent), joiners (17 percent), and targets (20 percent) have nearly identical rates of deployment. Drawing on the familiar 21-point Polity2 index of regime type (where +10 is a full democracy and −10 is a full authoritarian regime),[11] I find that belligerents using blocking units are slightly less democratic than their nonusing counterparts (−4.12 versus −2.62 for the 1800–2011 period), though the difference is only statistically significant in the modern era. When a democracy dummy variable (where Pol2 values ≥7) is used, however, a marked difference is observed: democracies represent only about 7 percent of states that used blocking detachments, compared with 18 percent of those that did not, for the 1800–2011 era. While democracies, including the Confederate States of America during the American Civil War and the short-lived First Republic of Venezuela during its War of Independence (1810–1812), have embraced such tactics, they are far more reluctant to do so than autocracies.[12]

More surprisingly, combatants with varying degrees of material power have adopted blocking units; they are not the sole preserve of the weak and desperate. Less powerful states certainly number among those adopting blocking detachments: Entre Ríos Province, during the Platine War (1851–1852) with the Argentine Confederation, and Siam, during its 1827 war with Vientiane, used battlefield executions

[11] Jaggers and Gurr (2004).

[12] This reluctance may have little to do with innate democratic values. Democracies typically fight much farther from their borders than autocracies – often far overseas – reducing the threat of desertion and defection and thus attenuating the need for blocking detachments. Note, too, that Reiter and Stam (2002, 73) find little statistical support for the proposition that democratic armies possess higher levels of morale.

and barrier troops to hold their forces together against superior foes. Yet vast empires have also employed blocking detachments, including the Ottoman Empire, Rabah Empire, the Kingdom of Dahomey, and Kokand. And, as noted above, even Great Powers, including China during the Korean War and Tsarist Russia on the Eastern Front of World War I, have relied on such units to round up deserters and to drive home frontal assaults. In fact, these data indicate that *stronger* powers or coalitions, as measured by relative share of total soldiers deployed during a war, utilized blocking detachments at a higher clip than their numerically inferior opponents.[13]

Soldier Motivation, Unit Cohesion, and the Trade-offs of Blocking Detachments

Blocking detachments are one possible solution to a more general problem: how to motivate soldiers to fight and die. At present, the existing literature has offered three families of explanations for soldier motivation and overall unit cohesion.

Some contend, for example, that soldiers are driven by ideology, especially nationalism, and thus fight to defend or advance a particular cause.[14] Others suggest that soldier motivation is tied to material benefits and selective incentives, including salaries and opportunities for battlefield spoils.[15] Perhaps the most widely cited argument centers around the role of primary group bonds: soldiers fight not for sweeping ideological visions or crass material gains but instead for each other. Strong emotional ties and shared sacrifices bind soldiers into bands of brothers that fight doggedly for fellow soldiers until extreme losses shatter the primary group.[16]

By comparison, scholars have been largely silent on how coercion might motivate soldiers and, in turn, undergird unit cohesion. Our theories tacitly assume, for example, that negative inducements are weaker than positive incentives. To cite one well-known study, Margaret Levi's consent-based approach to the study of military service

[13] This may indicate that larger armies suffer a greater probability of principal-agent problems that are "solved" using blocking units.

[14] Posen (1993); Levi (1997); Reiter (2007); Castillo (2014).

[15] Lichbach (1998); Weinstein (2007).

[16] Marshall (1947); Shils and Janowitz (1948); Stouffer et al. (1949); Henderson (1985); Stewart (1991); Wood (2003).

largely excludes the possibility that states coerce their citizens into fighting.[17] While acknowledging that coercion may play a role in motivating soldiers, Jasen Castillo notes that "fear alone cannot keep soldiers fighting." Eventually, "soldiers will begin to fear the enemy more than the blocking formations pushing them forward."[18] In a sweeping study of American soldiers since 1776, Christopher Hamner suggests that the practical realities of the battlefield render punishment an ineffective and unreliable instrument for motivation, leading states to embrace other more positive means of bolstering unit morale.[19] More generally, it appears commonsensical that soldiers will battle harder, and units hold together longer, when they are fighting for a cause rather than to escape punishment.

This omission leaves us without a proper accounting of coercion's relative effectiveness and associated trade-offs. It is time, as Hew Strachan has argued, "to take cognizance of the possibly positive consequences of punitive procedures for combat motivation."[20] Indeed, the possibility that states not only resort to these measures frequently but are rational to do so, and that both democracies and autocracies have drawn on such practices historically, needs to be entertained. Ideological appeals may find little purchase among certain soldiers, for example, while primary group bonds can actually facilitate desertion and defection, eroding rather than enhancing military effectiveness.[21]

Blocking detachments can therefore be a viable, perhaps dominant, strategy for generating combat power for many combatants. There are, however, multiple trade-offs inherent in their use, forcing states to privilege some elements of military effectiveness while accepting diminished performance in other areas. I follow the volume's shared definition of military effectiveness: "Militaries are effective to the degree that they can accomplish at acceptable costs the goals assigned to them by political leaders."[22] I examine the presence of trade-offs at the warfighting level, which I define as the tactical and operational levels of analysis, where tactics are associated with small unit maneuvers while operations encompass army-size maneuvers across at least one front.

Blocking detachments are associated with at least one central trade-off. Their presence can bolster a military's "staying power" – namely, its ability to absorb heavy losses while maintaining cohesion – but at

[17] Levi (1997). [18] Castillo (2014, 24). [19] Hamner (2011, 3).
[20] Strachan (1997). [21] Lyall (2016a). [22] See Chapter 1 of this volume.

extremely high cost, typically measured in terms of relative casualties suffered but also in lost matériel and missed battlefield opportunities. Concerns about staying power are often associated with combatants that are losing wars,[23] but this is not always the case; states can initiate wars and still rely on blocking detachments to secure their ambitions if unexpected casualties are incurred or attack momentum has been blunted. These detachments can buy time for regrouping by holding savaged units together and preventing further desertion or defection. Doing so, however, will worsen loss–exchange ratios, as soldier executions or imprisonment, along with tactical and operational vulnerabilities, conspire to increase losses and decrease capabilities for inflicting them. While these losses may appear favorable compared to complete military collapse, they are nonetheless worse compared to the counterfactual case of a similar combatant that did not need to deploy blocking detachments.

There are at least four specific trade-offs underpinning this "macro" trade-off between combat power and cohesion. First, blocking detachments pit cohesion against casualties: they can mitigate the risk of desertion, defection, and more diffuse foot-dragging by the threat or application of violence, but at the cost of self-inflicted casualties. Second, blocking detachments help militaries maintain tighter battlefield control over their officers and soldiers. At the same time, however, they reduce tactical and operational flexibility, again driving up casualties as militaries come to rely on rigid, costly, and less innovative practices such as frontal assaults. Third, blocking detachments reduce training demands, lowering the investment in human capital necessary to generate combat power quickly. Such gains come at the cost of lowering the skill level of soldiers, in turn reducing their ability to master their weapons and to execute complicated tactics and operations, including combined arms integration and movement under fire. Fourth, blocking detachments may stiffen the resolve of units, leading them to fight harder and longer than they otherwise would, but at the cost of creating new grievances, including the opening of a second front between officers and soldiers and soldiers and their blocking detachments. These trade-offs are summarized in Table 4.1.

Before exploring these trade-offs in greater detail below, it is worth noting that assessments of trade-offs hinge on identifying the

[23] Castillo (2014).

Table 4.1 *Coercion and the War-Fighting Trade-offs of Blocking Detachments*

Trade-off	Benefit	Drawback
Discipline vs. casualties	Decreased desertion, defection, shirking	Increased (self-inflicted) casualties
Tighter battlefield control vs. reduced flexibility	Enforce control (incl. over officers)	Rigid tactics and operational art
Reduced training demand vs. lowered skill	Substitute for training	Reduced weapons and tactical proficiency
Stiffened soldier resolve vs. new grievances	Reduced panics, increased fire volume	Worsened morale, open "second front"
Sum	Improved battlefield staying power	Worsened loss–exchange ratios

appropriate comparison group. Trade-offs can be identified in two ways: (1) the belligerent's own battlefield performance during a given war, where the positive and negative effects of blocking detachment within the same military across different facets that comprise military effectiveness can be isolated (a "within-case" comparison), and (2) the comparison to a similar belligerent that did not deploy blocking detachments (a "paired-case" comparison). This second comparison is helpful in establishing the counterfactual.[24] That is, how much worse (or better) would the state's battlefield performance have been had it not used blocking detachments? I use both types of comparison below.

Discipline versus Casualties

Blocking units represent an institutional response to the twin threats of desertion and defection that have plagued armies for centuries, particularly those armies that find themselves suffering high casualty rates at the war's outset. *Desertion* is defined as the unauthorized wartime withdrawal of a unit (or group of soldiers) from the battlefield or its rear areas with the intention of permanently abandoning the fight. Withdrawing from the war effort can take two forms: hiding from state

[24] Rosenbaum (2010); Rubin (2006).

authorities among the civilian population in an attempt to return to a prewar life or resorting to brigandage in rear areas without coordinating with enemy forces.[25] *Defection*, by contrast, is defined as the transfer of allegiance to the opposing side with the intention of taking up arms against one's former government.

If successful, blocking units may significantly reduce manpower drain – perhaps eliminating desertion and defection entirely – while also preventing the transfer of weapons and intelligence to one's opponent. Stemming this loss of manpower also has downstream benefits: preventing desertion undercuts the formation of brigandage units by deserters that prey upon logistical lines or local populations for food, weapons, and money. During the First World War, the Ottoman Army, for example, experienced massive desertion from its Greek, Armenian, and Arab soldiers, who, in turn, formed groups that attacked railway lines to pilfer supplies, hobbling Ottoman logistics.[26]

Similarly, blocking detachments not only restore a military's manpower but also prevent large groups of disgruntled, armed soldiers from heading home with the intent of toppling the regime itself. Shaky regimes may turn to blocking detachments as a battlefield form of "coup-proofing" that minimizes the chances that military indiscipline and desertion could become existential challenges to regime survival. There is a now-extensive literature on coup-proofing, particularly in Arab authoritarian states, where militaries are deliberately hamstrung – including circumscribed training, prohibitions on live-fire exercises, recruitment and promotion that favors certain (loyal) groups over merit – to reduce coup threat. These actions invariably trade regime security for military effectiveness, however.[27] And while the impetus behind the deployment of blocking detachments is to improve staying power, and the odds of eventual victory, by sealing soldiers in place, it is plausible that these measures can have spillover effects that contribute to regime security.

This externally imposed discipline comes with a marked downside, of course. In many cases, these units can inflict tremendous casualties on their own forces. Santiago Marino, a key leader of the (Second) Republic of Venezuela during its War of Independence (1812–1814), resorted to executing every fifth deserter caught by his special disciplinary formations. Given the small size of his army (less than five

[25] Lyall (2016b). [26] Beşikçi (2012, 270–272).
[27] Quinlivan (1999); Pilster and Böhmelt (2011, 2012); Talmadge (2015).

thousand soldiers at some points), the effect was enormous.[28] Desertion was also a widespread, chronic issue facing Iraqi forces throughout 1980–1988 Iran–Iraq war; according to one estimate, some twenty-five thousand soldiers deserted in a four-month period between December 1986 and March 1987 alone. Saddam Hussein would order his disciplinary units to execute thousands, and perhaps tens of thousands, of Iraqi soldiers for desertion over the course of the war; as early as May 1982, orders were sent to execute hundreds of soldiers to forestall rising panic within the ranks.[29]

There is also a hidden side to these casualties. Anecdotally, the rise of self-inflicted mutilation and maiming by soldiers desperate to escape both the battlefield and the wrath of blocking units is often noted. These behaviors reduce the available pool of recruits and, equally as important, contribute to the erosion of morale among remaining soldiers. Given these attempts at evasion, it is unsurprising that some blocking detachments, including those employed by the Soviet Union, had specific orders to comb field hospitals for soldiers suspected of deliberately harming themselves.[30]

Moreover, these units are costly to deploy since they represent a diversion of the best, and most loyal, soldiers away from frontline duties. This is a gamble that may not pay dividends: while desertion may be stemmed, the army's ability to inflict punishment on enemy forces will also be reduced by pulling these elite soldiers from the front. This is especially likely if blocking detachments become substitutes for proper training, thereby reducing soldiers' weapons proficiency and worsening their ability to fight. In addition, the psychological costs of serving as blocking forces can also negatively affect unit cohesion. Though we have few memoirs from participants in blocking detachments, military authorities have often provided special incentives to motivate these soldiers to continue their task. These incentives have ranged from higher pay and better rations and weapons to sanctioned battlefield looting. President Bashar al-Assad's quasi-blocking force, the National Defense Forces, has been granted preferential access to food, for example, along with official permission to loot from the battlefield.[31]

[28] Slatta (2003, 105). [29] Murray and Woods (2014, 176, 297).

[30] Ellis (2011, 330n59).

[31] More generally, the creation and staffing of these units can also represent a sizable diversion of resources away from the battlefield. Imperial China and the Soviet Union each supported their blocking detachments with substantial

Militaries adopting blocking detachments are essentially gambling that they can generate a credible deterrent to desertion and defection without inflicting enough casualties to sabotage the entire war effort. The exact location of this trade-off is an empirical question, one that depends on the detachment's size, its lethality, the extent of its bureaucratic apparatus, and the salience of the underlying grievances leading soldiers to challenge authority. That said, the trade-off between cohesion and casualties is a basic feature of blocking detachments, one that all militaries employing such units will inevitably face on the battlefield.

Control versus Flexibility

Blocking detachments can improve command and control (C2) in two ways. Tighter control can be exercised over officers if these units have the authority to punish them for failure. In turn, officers can use the presence of these units to threaten their soldiers with punishment, reinforcing their own control over the rank and file. Together, these pressures on soldiers and their commanders can improve battlefield C2, particularly when battlefield losses are piling up and when senior military commanders suspect disloyalty among their officers or soldiers. The result is a more unified effort within and across units, while also increasing the odds that orders will be carried out, improving battlefield performance somewhat when compared to a baseline of total C2 breakdown. Improved C2 also reduces the likelihood of a successful officer-led challenge, adding another layer of battlefield coup-proofing.

By extension, blocking units can also be deployed to enforce control over unreliable proxy forces. During the ongoing war in Ukraine, Russian forces have been accompanied by the Ministry of Interior's Dzerzhinskiy Division to serve as barrier troops behind Russian and rebel (proxy) lines. Reports place the Dzerzhinskiy Division near Debaltseve (in northern rebel controlled territory) and Mariupol, for example. Eyewitnesses have recorded at least five instances of punitive action taken against Russian soldiers by this Division; rebels, too, have been sanctioned.[32]

bureaucracies, rear prison camps, and extensive surveillance efforts that demanded tens of thousands of soldiers.

32 Sutyagin (2015, 9).

Tighter control in turn restores some options at the operational level, at least compared to militaries facing disintegration. Blocking detachments may reopen avenues of action previously ruled out by the threat of (further) soldier indiscipline. These stopgap measures, though inefficient, permit a combatant to launch operations that states in similar situations but without blocking detachments could not undertake. In fact, the threat of sanction exhorts units to prosecute riskier and costlier operations than otherwise possible, perhaps gaining a significant edge over adversaries that cannot muster similar efforts. Blocking units can, for example, drive near-suicidal frontal attacks that can swamp an enemy's defenses. Pushed past their natural breaking points, these units become key assets in an attritional struggle to grind a superior opponent down.

Reliance on these blocking detachments does reduce battlefield flexibility, however. In particular, these units can impose sometimes severe restrictions on the tactics and operational practices of fielded armies. Since blocking detachments substitute coercion for soldier initiative, militaries will often simplify their tactics and operations to limit their complexity for soldiers who are poorly motivated and trained (see below). Frontal assaults are a hallmark of forces with blocking detachments since they concentrate manpower at a fixed point, reducing the need for complicated coordination while allowing advancing units to remain tethered to their minders.

The need to maintain close physical proximity to blocking detachments also undercuts the ability of attacking armies to seize opportunities, especially during exploitation operations after breaking through enemy positions. Tactics and operational art will also innovate at a slower pace since many battlefield problems can be "solved" simply by throwing more soldiers into the fray, a luxury that armies not backstopped by these detachments might not be able to consider. These conservative tactics and operations are rational from the commander's point of view: they offer less risk of failure than bold but unproven operations. Better to muddle through and "only" lose men than risk personal sanction for abandoning orthodox battlefield practices.

Yet rigid command and control, along with simplified tactics and an unwillingness to rely upon individual soldier initiative, is a recipe for increased casualties. Conservative tactics and the need to advance at the pace of blocking detachments create new vulnerabilities that enemies can exploit to inflict greater casualties than otherwise possible.

These same vulnerabilities also limit the amount of damage done to an enemy; blocking detachments may slow rates of advancement, for example, providing scattered enemy forces time to regroup and escape encirclement. By slowing rates of innovation, blocking detachments impose constraints on killing proficiency, forcing belligerents to miss out on the early adoption of alternative tactics or operations that might improve loss–exchange ratios.

The presence of blocking detachments also introduces C2 vulnerabilities. If these units are required to maintain military cohesion, then they become high priority targets. Given their distinctive profile on the battlefield, these units are uniquely vulnerable to counter-C2 efforts. This is especially true in the modern era of warfare; these units can be found, fixed, and targeted by aircraft, drones, and electronic warfare to degrade their capabilities. Cracking these disciplinary units creates new opportunities for desertion and defection from previously bottled up soldiers. At a minimum, targeting blocking detachments will reduce an army's battlefield flexibility by ruling out certain tactics or operations that exposes these units to discovery. Targeting these detachments also reduces their mobility by making it more difficult for them to move around the battle space, lengthening reaction times to setbacks that threaten to turn into routs. For example, American aircraft specifically sought and targeted the Special Republican Guard, Special Police, and regular Republican Guards, all of which deployed behind frontline Iraqi units with instructions to shoot deserters during the 2003 war. Degrading these units facilitated the desertion of tens of thousands of Iraqi conscripts.

Reduced Training versus Lower Skill

Another key aspect of creating power, if an unglamorous one, is training. Realistic training exercises are thought essential to fostering cohesion within units and imparting the skills necessary to wage combat, particularly in the modern era of sophisticated weaponry and complex combined arms operations. Training provides the opportunity to engrain tactical thinking and master new technologies.[33] It acts as a key mechanism of socialization in which civilians are transformed into professional soldiers, imbibing the military ethos while forming primary

[33] Strachan (2006); Talmadge (2015, 13).

group bonds through shared hardships that create strong unit cohesion. Realistic training also increases military effectiveness by creating a familiarity with the noise and confusion of the modern battlefield, reducing the destructive nature of surprises that can shatter cohesion while also socializing soldiers into the habit of firing their weapons, something they typically are reluctant to do in the absence of prewar training.[34]

States may lack the resources to devote to realistic training. They may also fear the consequences of diffusing prewar weapons training throughout their population given suspected disloyalties. Once war begins, states may lack the time to provide realistic training, especially if battlefield casualties are heavy. In these situations, states need to mobilize large numbers of soldiers quickly and have them make an immediate impact, even if they are unskilled. Blocking detachments provide one solution: coercion allows states to shortchange training, sacrificing quality for quantity and using threatened violence rather than socialization as the glue holding units together. This solution provides a framework for action that does not hinge on a high level of skill or training but that generates combat power quickly, if crudely, a kind of exoskeleton for low-skill soldiers.[35]

In fact, by halting widespread desertion or panic, blocking detachments can actually buy time for more sustained training over the medium to long term. The presence of these units will also improve killing proficiency by forcing soldiers to overcome their reluctance to fire their weapons. This is especially true in the pre–World War I era, where the absence of rigorous (prewar) training meant that soldiers still had a strong aversion to killing.[36] Units with blocking detachments will be more proficient – as measured by volume of fire – than comparable units without blocking units, though perhaps not as effective as units that never needed these detachments in the first place.

Of course, the substitution of coercion for sustained training imposes costs. Armies with blocking detachments will muster and deploy

34 Collins (2008, 49).

35 This solution, however, can create a vicious cycle where low-skill soldiers force belligerents to deploy blocking detachments, which in turn further reduce demand for skilled soldiers. Blocking detachments also contribute to this deskilling if they draw their personnel from the best-trained units while also creating incentives to adopt simpler tactics and operations.

36 Grossman (1996).

soldiers with lower-than-average skills relative to their nonblocked opponents. Tactical skills are an important, if often overlooked, aspect of military effectiveness: they help soldiers maximize their use of terrain, reducing their exposure to enemy fire while maximizing their ability to inflict casualties.[37] Weak skills translate into reduced tactical proficiency, imposing constraints not only on basic tactics but also the ability to carry out complicated operations that require a high degree of coordination. Unskilled soldiers will also have reduced means to seize sudden battlefield opportunities or to improvise within their commanders' guidelines in order to stay alive and inflict casualties.

The result is again a ballooning of the costs incurred by combatants utilizing blocking detachments: casualties will be higher than nonblocked armies and loss–exchange ratios far worse as less skilled soldiers fare worse on the battlefield, reducing their ability to inflict casualties even as their own losses mount. Reduced firearm proficiency will also result in a decreased ability to kill enemy soldiers, resulting in worsened loss–exchange ratios. Reduced opportunities to acquire skills also extend to more prosaic issues such as maintenance: without these skills, armies risk the interruption of their operations due to logistical delays in repairing and resupplying matériel.

Resolve versus Grievance

Perhaps the most intuitive reason for using blocking detachments is that coercion can increase soldier resolve, especially among units with poor morale and indiscipline but that have not (yet) committed mass desertion or defection. By foreclosing the ability to retreat, blocking units may force soldiers to fight harder and to absorb higher casualties than they otherwise might have if left to their own devices. These units can be especially valuable when enforcing discipline and order among soldiers who view the state as illegitimate or are in danger of breaking from heavy battlefield casualties. In these instances, blocking units represent a last-ditch effort to force soldiers to fight, avoid costly retreats, and to buy time for reinforcements to arrive.

The effects of these blocking units can also extend beyond their immediate surroundings to influence other (nonblocked) units. The deployment of blocking units in select circumstances may create a

[37] Biddle (2007).

demonstration effect via indirect deterrence that persuades other units to shape up to avoid similar sanction. By shoring up resolve in wavering units, blocking detachments can stop the spread of panic to other units. Stephen Rosen, for example, argues that informational cascades lead steadfast soldiers (or units) to desert because they witness others doing so, leading them to reassess their own positions as the preferences of others are revealed. Units may wish to stand fast but will nonetheless collapse into disarray as their soldiers update their expectations about the likelihood of others' desertion and the probability of winning the war.[38]

Yet while coercion can induce resolve mechanically, its use only creates further grievances among soldiers and officers. If soldiers were reluctant to fight on behalf of the regime before, threatened violence is unlikely to generate genuine pro-regime sentiment. Instead, the use of blocking detachments will be treated as a credible signal that the regime believes its soldiers are unreliable, and that they will exercise less initiative (unless organizing desertion) and less resolve if left unguarded. These grievances may lead to catastrophic spirals of desertion and defection if a unit is separated from its blocking detachment during battle, especially if soldiers were from targeted groups that had experienced violence or discrimination at the hands of the regime in the prewar era. Blocking detachments can reinforce rather than suppress existing ethnic and other fault lines within these units if their punishment is applied unevenly. In turn, these grievances may be exploited by opponents' propaganda calling attention to the graphic nature of the regime's own devaluation of their lives.

More generally, grievance formation can spark an intramilitary "second front" pitting soldiers against their officers and units against their blocking detachments. The creation and deployment of blocking units can drive a wedge between officers and soldiers by setting up dueling incentive structures. Officers will be punished by blocking detachments for not pushing their soldiers forward, creating incentives to use them callously to avoid their own execution or cashiering for poor performance. Soldiers will naturally resent these operations and are liable to strike back by "fragging" officers. Bonds of trust between officers and their soldiers will also crumble, hobbling coordination and strategy. At its extreme, this resentment may push soldiers to organize

[38] Rosen (2005, 125).

collective action against the blocking detachments themselves, either by purposely attempting to separate during battle or even by turning their weapons against them.

Trade-offs in Action: The Soviet Experience at Stalingrad and Kursk

Operation Barbarossa, Hitler's plan for invading the Soviet Union, was launched on June 22, 1941, and quickly inflicted staggering losses on Soviet forces. Enormous casualties – some three million Soviet soldiers were captured as prisoners of war in the first six months alone[39] – and widespread desertion and defection pushed the Red Army to its breaking point. Despite local successes in blunting German momentum, as at Moscow, an increasingly desperate Red Army instituted blocking detachments as official policy in 1942.[40]

I use process tracing to explore the trade-offs associated with Soviet blocking detachments during key battles at Stalingrad (August 23, 1942, to February 2, 1943) and Kursk (July 5–August 23, 1943). The Soviet campaign represents one of the largest deployments of blocking detachments in warfare – and one of the most lethal – casting the trade-offs associated with their use in sharp relief. The Eastern Front as a whole represents an "edge case" in the history of war, both for its overall lethality and high intensity, and provides the opportunity to examine Soviet blocking detachments during offensive and defensive actions. These battles, while not exhaustive, help generate propositions about trade-offs that can be tested comparatively in other settings.

Soviet commanders had already implemented coercive measures in an ad hoc fashion as early as June 1941 to hold units together against the German onslaught. Deserters and defectors were often shot on the spot during summer 1941 to stiffen resolve; in some cases, whole units turned against their political officers (*politruki*) in a bid to escape punishment or the war itself.[41]

In at least one notable instance, Soviet artillery fire was called down on a unit that was defecting to the Germans at the Battle of Chernevo

[39] Snyder (2010, 175).

[40] The literature on the Eastern Front is vast. Key works include Glantz (2014); Citino (2007, 2012); Reese (2011); Merridale (2006); Glantz (2005a, 2005b); Glantz and House (1999); Erickson (1999a, 1999b).

[41] Reese (2011, 168).

in 1941.[42] On August 16, 1941, Order No. 270 was issued by Stalin, which extended authority to commanders to shoot deserters and arrest their family members; it also prohibited encircled soldiers from surrendering and authorized the battlefield cashiering of hesitant, incompetent, or just plain unlucky, officers.

These measures failed to stem the tide of desertion and battlefield setbacks, however. Stalin then resorted to more drastic measures, personally writing Order No. 227 on July 28, 1942.[43] The order called for the immediate creation of two-hundred-man blocking detachments to be staffed by Red Army soldiers and positioned in the rear of all units. They were part of a broader interlocking monitoring and sanctioning system of secret police (NKVD), surveillance and censorship, and penal battalions for officers (*shtrafbaty*) and companies for soldiers (*shtrafroty*).[44] Blocking detachments were to establish positions two to three kilometers behind the front lines to fulfill three tasks: return straggling or lost soldiers to their units; prevent desertion and defection through fear of sanction; and imprison or kill commanders or soldiers that deserted, defected, exercised poor judgment, or otherwise represented a threat to cohesion.

This harsh measure was justified, Stalin believed, by continued indiscipline:

> What do we lack? There is no order and discipline in companies, battalions, regiments, in tank units and air squadrons. This is our main deficiency. We should establish in our army the most stringent order and solid discipline, if we want to salvage the situation, and to keep our Motherland... Panic-mongers and cowards should be wiped out on the spot.[45]

The adoption of these blocking detachments was not due solely to battlefield losses, however. Stalin and his High Command (Stavka) faced a deeper structural problem: allegiance to the regime was highly variable among soldiers, particularly those drawn from ethnic groups that had suffered from repressive prewar Soviet nation building and collectivization drives. These soldiers had little desire to fight for a regime that had victimized their families, helping to explain why desertion and

[42] Maslov (1996, 428–429). [43] Khlevniuk (2015, 223).

[44] Ideological changes, including patriotic and nationalist appeals, were also adopted. See Brandenburger (2002).

[45] Order No. 227, People's Commissar of Defense of the USSR, July 28, 1942, Moscow.

defection was so high among Soviet forces during 1941–1942. In addition, many of these soldiers, especially Ukrainians and Belorussians, had homes in now German-occupied territories, creating both motive and opportunity for soldiers to slip away from the ranks. Many Soviet commanders viewed their own soldiers with suspicion, accusing them of divided loyalties and an unwillingness to bear the necessary costs to protect the Soviet regime. Nor could commanders rely on primary group bonds to instill discipline and maintain cohesion: by one estimate, loss rates were so high during these initial years that the average frontline tour for an infantryman before death or serious wounding was only three weeks.[46]

Yet reliable information about these detachments, despite their well-known role, remains difficult to obtain, in part because of prior censorship and current sensitivities; Order No. 227 itself was not publicly released until 1988. Most scholarly treatments of the Eastern Front typically marginalize the effects of these units, pausing only to highlight their shocking nature before returning to blow-by-blow accounts of various battles. In fact, the best accounts of these units are actually found in Soviet and post-Soviet literature.[47]

That said, Soviet-era archives have gradually revealed their secrets about blocking detachments, and even more material is slated to be declassified beginning in 2017.[48] These archival records, if still incomplete, have sparked a vigorous debate how, and if, blocking detachments affected Soviet military effectiveness. At one extreme, nationalist Russian historians have questioned the very existence of most of these units, suggesting that their role has been exaggerated by Western historians seeking to denigrate Soviet contributions in Nazi Germany's defeat (*degeroisatziya*).[49] Other historians have argued that executions were so infrequently and haphazardly carried out that they did not serve as a credible deterrent. "Soldiers may have been afraid," Roger Reese has argued, but "that does not explain the compliance of the majority of the army."[50]

[46] Merridale (2006, 16).
[47] See, e.g., Emmanuil Kazakevich's *Star* (1946) and *Two Men on the Steppe* (1948), Grigorii Baklanov's *South of the Main Blow* (1958) and *The Dead Feel No Pain* (1966), and Yuri Bondarev's *The Battalions Request Fire Support* (1957). For the post-Soviet era, see Ellis (2011, 38–39).
[48] Statiev (2010, 722n1).
[49] For one such account, see Starikov (2014, 120–124).
[50] Reese (2011, 173).

Others assign a much more prominent role to these units, however. "All soldiers shared some measure of fear," Catherine Merridale has written, and "the NKVD soldier with his pistol, shooting stragglers in the back, is an abiding image of this war."[51] Alexander Statiev has reached a similar finding, noting the "nearly unanimous opinion of Soviet veterans [that] the threat to be sent to a penal unit strengthen[ed] discipline."[52] Perhaps the strongest claims are made by David Glantz in his monumental study of Soviet military performance: "The iron discipline and administered by Stalin . . . served as the essential 'glue' that bound the Red Army together as a coherent fighting force and permitted it to survive and, ultimately, prevail despite the appalling combat conditions its soldiers had to endure."[53] A consortium of Russian military historians has echoed this view of blocking detachments as indispensable. Order No. 227, they concluded, played "a major role in increasing the resilience and military activity of Soviet forces [and] in creating a turning point in the course of military operations."[54]

The Cases: Stalingrad and Kursk

The Battle of Stalingrad is often considered a turning point in Nazi and Soviet fortunes on the Eastern Front. The battle began with Germany's Army Group South launching its massive Operation Blau to shatter Soviet forces at Stalingrad, which occupied a key defensive salient barring further Nazi advances. Soviet forces bent but did not break, blunting German momentum before launching their own Operation Uranus (November 19, 1942), a two-pronged attack aimed first at weaker Romanian and Hungarian units protecting the Sixth Army's flanks before encircling the entire Sixth Army. Its surrender in February 1943 represented the first time that Soviet forces had encircled and destroyed a German Army. The cost was high for both sides, as the battle degenerated into brutal house-to-house fighting within Stalingrad's shattered environs. Soviet forces lost an estimated 1,129,619 casualties; Nazi forces, about 850,000 soldiers killed, wounded, or captured. Soviet matériel losses were also enormous: some 4,341 tanks, 15,728 artillery pieces, and 2,769 aircraft were destroyed; the Nazis lost 500 tanks, 6,000 artillery pieces, and 900 aircraft.[55]

[51] Merridale (2006, 317). [52] Statiev (2010, 745). [53] Glantz (2005b, 582).
[54] Zolotarev (1996, 330).
[55] Estimates are from Krivosheev (2009, 107–109, 221–222).

If Stalingrad blunted German momentum, then Kursk represents the watershed mark when strategic initiative passed into Soviet hands for good. Desperate to restart its stalled campaign, Germany launched Operation Citadel, a massive effort designed to punch through a large concentration of Soviet forces in the Kursk salient. The German offensive envisaged a complicated double pincer maneuver that would encircle and then destroy trapped several Soviet armies on the Voronezh and Central Fronts. Yet Soviet commanders, armed with intelligence intercepts, had constructed a massive defense-in-depth that, for the first time, blunted a blitzkrieg-style offensive before it reached Soviet strategic depths. The Nazi thrust only penetrated about 8–12 kilometers in the north and up to 35 kilometers in the south, before bogging down amid the eight defensive belts that extended up to 170 kilometers behind Soviet front lines. Soviets forces shifted to the offensive on July 12, launching two operations (Kutuzov and Polkovodets Rumyantsev). The fighting, which witnessed some of the largest tank battles in history (as at Prokhorovka, July 12, 1943), was extremely bloody. Some Soviet units essentially disintegrated in place, sustaining up to 70 percent losses in a desperate bid to prevent Nazi incursions.[56] Overall, 1,041,150 Soviet soldiers were killed, wounded, captured, or missing, while over 6,000 tanks and assault guns were destroyed or rendered nonoperational; Axis forces lost 252,182 soldiers, along with 760 tanks and assault guns.[57]

Space constraints do not permit a full examination of these battles, each of which has its own (vast) dedicated literature. Taken together, however, these battles offer a plausibility probe of the trade-offs associated with blocking detachments in various phases of modern, high-intensity warfare. Blocking detachments were partially deployed at Stalingrad, for example, but had proliferated to nearly every unit at Kursk. Soviet forces were both on the defensive and offensive in these battles, and in each case transitioned from defensive positions to counteroffensives, permitting investigation of blocking detachments during various campaign phases. And these units were deployed during different types of combat, ranging from the close quarters, block-to-block fighting of Stalingrad to mobile operations, including some of the largest combined arms operations ever undertaken.

[56] Glantz and House (1999, 275).

[57] Estimates are from Krivosheev (2009, 123–124, 228).

Cohesion versus Casualties

The battles at Stalingrad and Kursk illustrate the greatest asset of blocking detachments: their ability to increase unit cohesion mechanically by foreclosing opportunities for desertion and defection. A staggering number of soldiers were detained and arrested by the forty-one detachments (about eighty-two hundred men) operating on the Stalingrad and Don Fronts. Some 51,728 soldiers, or 37 percent of total Red Army soldiers deployed on these fronts, were detained for suspected desertion or unauthorized leave from their units.[58] Between August 1 and October 15, 1942, an estimated 140,755 soldiers were detained by 193 blocking detachments (38,600 soldiers) across all fronts. While only partial records are available, a similar story emerges at Kursk: blocking detachments stationed behind three of the armies caught almost seven thousand soldiers from July 8 to 14 alone, more than half stemming from eight different supposedly elite Guards divisions. The Ninety-Third Guards Rifle lost 10 percent of its strength ($n = 969$), for example, while a further 734 soldiers were detained on July 8 alone from a single unit after its panicked flight from Nazi forces.[59]

These blocking detachments functioned as a catch-and-release program that steered the majority of wayward soldiers – stragglers, those absent without authorization, malingerers – back to their units. While filtration points acted to channel those suspected of desertion or defection into prison camps or public executions, the majority of soldiers went right back to their units, boosting Soviet manpower enormously. During 1942–1943, 1.25 million soldiers were caught away from units without proper documentation and another two hundred thousand were identified as stragglers. In 1943, the immediate rear area (extending twenty-five kilometers behind the front line), the NKVD detained 158,585 soldiers for "straggling behind," another 42,807 men for unauthorized departure, and 23,418 for suspected desertion. Another eighteen thousand soldiers were sent to penal units or the Gulag.[60] If nothing else, these detachments prevented a massive loss of Soviet

[58] "Spravka OO NKVD STF v UOO NKVD SSSR o deyatel'nosti zagraditel'nykh otryadov Stalingradskogo i Donskogo frontov [Ne ranee 15 oktyabrya] 1942 g.," in Pogonii (2000, 230–231). A total 36,109 soldiers were detained on the Don Front, and another 15,649 were detained at Stalingrad.

[59] Reese (2011, 174). [60] Reese (2011, 174–175).

personnel that could have turned into an exodus like that seen during the last days of Tsarist Russia's Army in 1917.

Given the scale and enduring nature of the problem, it is clear that blocking detachments did not completely solve the Red Army's cohesion problems. An estimated 4.4 million incidents of desertion, defection, and unauthorized leaves were recorded over the course of the war.[61] For comparison, the Red Army had about 6.7 million soldiers in the battlefield at the time of the Battle of Kursk. Large units, including whole companies and battalions, deserted at Stalingrad, and soldiers continued to try to reach German lines even after Operation Uranus was launched, including the entire Forty-Second Rifle Division (September 24, 1942).

Desertion would only ebb in late 1944, and firm conclusions about how much more (or less) desertion would have occurred had blocking detachments been absent are difficult to reach. That said, intriguing glimpses are on display: at Kursk, nineteen men of the 179th Penal Company of the 13th Army's 148th Rifle Division defected to the Germans, while others fled to the rear because a blocking detachment was not present (March 1943). This incident, and others like it, forced Soviet commanders to reissue orders explicitly stating that penal units could only be used in situations where blocking detachments were present to monitor them.[62]

The blocking detachments exacted a heavy price, though, in terms of increasing Soviet casualties. Red Army detachments and their NKVD counterparts executed an estimated 158,000 of their own soldiers.[63] While these executions represented "only" 3 percent of the total number of soldiers detained, these losses in an absolute sense were enormous, representing at least eleven full-strength Soviet divisions. Some 13,500 soldiers alone were shot at Stalingrad within the space of two weeks.[64] Put differently, these casualties rivaled total US Army losses in the European and Atlantic theaters (185,924) and dwarfed those suffered in the Pacific campaign (106,207).[65] An additional 450,000 soldiers were captured by blocking detachments and reassigned to penal battalions. These units conducted high-risk missions

[61] Krivosheev (2009, 41).
[62] "Direktiva voennogo soveta Tsentral'nogo fronta no.027 ot 18.4.43 g.," quoted in Glantz (2005, 578).
[63] Merridale (2006, 157). [64] Merridale (2006, 157).
[65] Statistical and Accounting Branch Office of the Adjutant General (1946).

such as route clearance for Soviet offensives in exchange for reduced criminal sentences; casualties were staggering, often reaching 50 percent or greater of unit strength in just one operation.[66] In short, blocking detachments reduced cohesion problems but did so at the direct cost of massive casualties.

Control versus Flexibility

Blocking detachments also enabled Soviet commanders to maintain a degree of control over their soldiers likely not possible in their absence. Their effects are visible in two ways. First, Stavka, fearful that its commanders had grown overly cautious after the defeats of 1941, viewed these units as mechanisms that prevented premature withdrawal by commanders paralyzed by the threat of encirclement. As such, blocking detachments removed an element of discretion from the commander's purview by foreclosing any unauthorized retreats and by emphasizing offensive action.[67]

Second, these units could enforce control over regular units that were suffering catastrophic losses, whether as part of an attritional defense-in-depth strategy or offensive operations. At Kursk, losses among frontline Soviet divisions ranged from a low of 20 percent to as high as 70 percent, casualty rates likely unsustainable in most, if not all, armies.[68] The Fifty-Second Guards Division, for example, was virtually annihilated in the first few days of the German offensive at Kursk, its soldiers sacrificed in an attritional effort to limit the depth of German penetration. Similarly, the threat of coercion pushed soldiers forward to seize the momentum from stalled German attacks even when units were seriously depleted. Soviet casualties spiked threefold, from seven thousand on July 11 to twenty thousand soldiers on July 12, 1943, when the Soviets pivoted from the defensive to offensive at Kursk, yet continued to push forward.

Yet reliance on blocking detachments to maintain control translated into reduced battlefield flexibility and thus indirectly contributed to higher casualties. Despite improvements over time, Soviet tactics and operational art remained fairly simple, if not crude, for much of the war. This is true of both Stalingrad and Kursk, where even the most

[66] Glantz (2005a, 577). [67] Reese (2011, 97).
[68] Glantz and House (1999, 275).

innovative commanders – often in armor units – continued to rely on frontal assaults and basic maneuvers rather than more complicated operational art such as mobile defense or double envelopment. As Glantz acknowledged, the stereotype of the Red Army as a "monolithic and rigid force that employed artless steamroller tactics to achieve victory regardless of cost" was largely correct. The Red Army sought to absorb German offensives and then shift to the attack once their momentum had been halted, moving "in painstakingly rigid fashion while on the offense, often artlessly and regardless of cost."[69]

Blocking detachments conspired to narrow battlefield flexibility by design: to remain a credible deterrent, these units had to remain tethered to regular line units. As a result, Soviet commanders embraced simplified, rigid tactics that assumed military cohesion would suffer if a gap emerged between regular units and their minders. The need to preserve this linkage created incentives to rely on costly frontal assaults where blocking detachments could maintain near constant surveillance. Similarly, exploitation efforts were curbed for fear that attacking units might become separated from their blocking detachments.

Moreover, the fact that blocking detachments could punish officers created reinforcing incentives to rely on tried-and-true, if costly, tactics and operations and to eschew more innovative but riskier approaches that might save lives. Fearing cashiering if seen as insufficiently aggressive, commanders launched many mistimed offensives that stalled out in part because they felt pressured to take action even if their men and matériel were depleted.

In short, blocking detachments restored some measure of control over units and their commanders, no minor feat given the Red Army's state of disarray in 1941–1942. Yet these blocking detachments also undercut battlefield flexibility, increasing the predictability and vulnerability of Soviet forces and indirectly contributing to greater casualties and worsened loss–exchange ratios. Yes, in one sense the blocking detachments *recovered* some measure of flexibility; some units, if not the entire Red Army itself, may have completely disintegrated without their presence, restoring the ability of commanders to wield combat power on the battlefield. But the rigid tactics and operational art dictated by the same presence of these units represented a loss of battlefield freedom relative to the options available to commanders who did not have to rely on coercion to maintain control. In this case, Soviet

[69] Glantz (2005b, 618).

coercion allowed German forces to punish the Red Army heavily for its inflexibility, inflicting casualties at a rate above the baseline expectation of losses had these units not been necessary.

Training versus Skill

Order No. 227 and the threat of sanction from blocking detachments also permitted Soviet commanders to create new units at a breakneck pace. Coercion became a substitute for realistic training; raw recruits often received as little as two weeks training before their assignment to the front.[70] In some cases, civilians were actually dragooned into service, while whole units were thrown together from remnants of shattered units. Under such conditions, lasting primary group bonds were unlikely to form, creating units with potential discipline problems, low initiative, and unskilled soldiers. Blocking detachments thus acted as cement to keep these units together, while also allowing commanders to shorten training cycles. Faced with short time horizons and severe casualties, Soviet commanders could not afford peacetime training regimens designed to create cohesion through socialization. Coercion offered the quickest pathway to generating the most combat power with the least amount of time and resources.

Predictably, shortening the training cycle meant that Soviet conscripts lacked the time to acquire necessary weapons proficiency and tactical skills before being thrust into battle. Soviet Generals at Stalingrad immediately recognized the trade-off and lamented the poor training of Soviet infantry. General Malinin, Army Chief of Staff and present at Stalingrad, complained that "our infantry are useless . . . The artillery is doing its job, keeping the enemy's heads down, but the infantry won't stand up and push forward."[71] General Zhadov, commander of the Sixty-Sixth Army of the Don Front, also emphasized how unskilled infantry were a liability in battle: "infantry, particularly the new divisions, are not trained, don't know how to fight, and aren't capable of carrying out their assigned tasks."[72]

Using blocking detachments also had a detrimental effect on the Red Army's overall skill level. Blocking detachments were authorized to

[70] Reese (2011, 138).

[71] "Dokladnaya zapiska OO NKVD DF v UOO NKVD SSSR o nastupatel'nykh operatsiiakh 66-i armii, 30 oktybriya 1942 g.," in Pogonii (2000, 253).

[72] "Dokladnaya zapiska OO NKVD DF v UOO NKVD SSSR o nastupatel'nykh operatsiiakh 66-i armii, 30 oktyabrya 1942 g.," in Pogonii (2000, 252).

capture suspected deserters among armor and aviation units and reassign them to penal battalions as infantry. Having pilots, tank drivers, and, crucially, maintenance crews serve instead as infantry led to a deskilling of the Red Army at preciously the moment when these specialties were most needed.[73] For example, Soviet armored formations sustained high attrition rates at Stalingrad – some units lost 80–90 percent of their tanks – due mostly to mechanical breakdowns rather than enemy fire. These mechanical deficiencies foiled Soviet efforts at exploiting German collapses during Operation Uranus, letting a key opportunity for deep penetration slip by.[74]

Resolve versus Grievance

By foreclosing most opportunities for soldiers to abandon their units, blocking detachments improved overall resolve, if only in a negative sense. Widespread panics were reduced once blocking detachments were implemented, especially when compared to the dark days of summer 1941, when desertion and defection occurred even among units not yet engaged in combat. Coercion thus offered a partial substitute for patriotism and Russian nationalism, especially among populations that had experienced prewar repression and for whom calls to support the regime rang hollow. Accurately gauging the volume of weapons fire, often a measure of unit resolve, is a difficult proposition, though it is clear that Soviet forces began inflicting a greater number of casualties on German forces once blocking detachments were deployed.

Perhaps more telling, however, is the tinge of desperation and gallows humor that accompanied frontline soldiers' accounts of their own resolve. As one popular song recounted:

> The first shell punctured my gas tank.
> I jumped out of the tank – how I do not know.
> So they called me over to the special section [the blocking detachment]:
> "Why didn't you burn up with your tank, you son of a bitch?!"
> So I answered, and this is what I said:
> "I'll be sure to do that the next time we attack."[75]

Still, bolstering resolve via coercion created its own problems. Even as late as Kursk, soldiers attempted evasion if blocking units failed to

[73] Statiev (2010, 729). [74] Glantz and House (2014, 605).
[75] Reese (2011, 175).

seal possible escape routes. Soldiers quickly came to resent their presence, speaking openly of a "Second Front" being organized against them.[76] The NKVD, which monitored soldier attitudes closely using a vast network of informers, was alarmed by the possibility that "hostile elements" might exploit the creation of blocking units to induce Soviet soldiers to desert or defect.[77] The specter of armed mutiny also worried the NKVD.[78] Officers often became focal points for these grievances since soldiers could not determine whether executions and reassignment to penal battalions were genuinely for restoring order or merely an insurance policy for the commander to highlight his "resolve."

Soldiers took particular exception to the arbitrariness of these punishments; in many cases, innocent soldiers were prosecuted or killed. Soldiers sometimes resorted to killing ("fragging") their own officers, creating a potential crisis for command and control.[79]

Discussion

As the Soviet experience demonstrates, blocking detachments can play a critical role in maintaining unit cohesion under punishing conditions. This battlefield staying power came at tremendous cost, however. Indeed, the combination of executions, officer fragging, soldier mutilation, and reduced tactical and operational flexibility induced by these units produced far more Soviet casualties and a poorer loss–exchange ratio than otherwise expected by strict calculations of efficiency.[80]

The Red Army's use of blocking detachments also highlights several additional trade-offs. The presence of these units likely contributed to the barbarization of warfare on the Eastern Front, for example.[81] To be sure, both Hitler and Stalin refused to adhere to existing agreements

[76] See, e.g., "Dokladnaya zapiska OO NKVD STF v UOO NKVD SSSR 'O reagirovaniyakh lichnogo sostava chastei i soedinenii na prikaz Stavki No. 227, 14/15 avgusta 1942 g.,'" in Pogonii (2000, 191).

[77] Pogonii (2000, 187).

[78] See, e.g., "Spetssoobshchenie OO NVKD STF v UOO NKVD SSSR 'Ob otritsatel'nykh vyskazyvaniyakh otdel'nykh voennosluzhashchikh Stalingradskogo fronta v svyazi s izdaniem prikaza Stavki No. 227, 19 avgusta 1942 g.,'" in Pogonii (2000, 190–192).

[79] Merridale (2006, 192). [80] Millett and Murray (1988, 3).

[81] Intriguingly, the work most centrally associated with this barbarization thesis – Omer Bartov's *The Eastern Front, 1941–45, German Troops and the Barbarization of Warfare* – omits any mention of Red Army or NKVD blocking detachments (Bartov 2001).

governing the treatment of prisoners of war. As Timothy Snyder points out, the first concentration camps to appear on the Eastern Front were designed specifically to hold Soviet prisoners of war captured during the encirclement battles of 1941.[82] Yet while Hitler worked feverishly to dismantle incentives for individual Soviet soldiers to consider surrender, Soviet blocking detachments did so from the rear, creating a kill-or-be-killed environment in which Soviet soldiers avenged themselves upon captured Germans. The combination of Nazi and Soviet policies created a setting where neither surrender nor retreat were viable options for most Soviet soldiers, contributing to battles where quarter was neither given nor accepted. The result was a savage reciprocity where Soviet and Nazi POWs faced staggering mortality rates and where scores of soldiers (on both sides) were summarily executed while trying to surrender.[83]

In addition, the use of blocking detachments generated political trade-offs. Red Army and NKVD blocking detachments had an extensive role in policing – or, more aptly, "reoccupying" – newly liberated areas in Ukraine, Belarus, the Northern Caucasus, and the Baltic states. Viewing these populations as disloyal and potentially restive, the Red Army deployed blocking detachments to suppress local populations. These clampdowns included forced population resettlement (as in Chechnya), the rooting out of suspected traitors and their networks, and the (re)conscription of soldiers who had taken to hiding among locals when their units were overrun during German offensives in 1941–1942. From the regime's point of view, blocking detachments had political benefits in the form of reconsolidating Soviet power; the trade-off, measured in terms of additional civilian lives destroyed, was judged as necessary and unavoidable.

Some belligerents, however, have concluded that the political costs associated with blocking detachments are too high to countenance their battlefield deployment. Even desperate belligerents suffering from poor cohesion and high desertion rates have delayed adopting blocking detachments out of concern for their political consequences. The Confederate States of America, for example, was faced with endemic desertion and a corresponding crippling manpower shortage but only belatedly authorized the creation of dedicated forces to hunt down and return deserters in 1863–1864.[84] Desertion varied sharply across (and

[82] Snyder (2010, 175–179). [83] Morrow (2014, 220–224).
[84] Weitz (2005, 193–195).

within) CSA states, forcing President Davis to temporize in his decision about introducing these units for fear of upsetting his political supporters. But as increasingly large bands of deserters began preying upon locals, he was forced to adopt drastic measures to curb the steady erosion of support for the war among victimized populations. Similarly, Syria's President Bashar al-Assad's appears to have resisted a full-fledged commitment to these forces until 2014, when a combination of battlefield setbacks, territorial losses, rampant desertion, and recruitment woes drove him to accept the political costs associated with relying on blocking units.[85]

Finally, Soviet practices also raise the key issue of whether these crosscutting war-fighting effects scale up to affect war outcomes. Of course, many factors contribute to a state's victory (or defeat), and it can be difficult to isolate the specific effects of any one policy or practice. This is especially the case with blocking detachments, where selection effects concerning the adoption of these units and timing of their implementation are wrapped up in broader war dynamics and belligerent characteristics. Was the belligerent already on the glide path to defeat when these units were created, thereby spuriously associating blocking detachments with defeat? Did these units delay the collapse of an army or actually hasten its demise? All of these scenarios are plausible, and so caution is warranted when assigning blame and credit to blocking detachments.

That said, there are a sufficient number of historical examples of belligerents deploying these units and winning major wars – the Bolsheviks during the Russian Civil War, the USSR during the Winter War (1939–1940) and Eastern Front, the Chinese government during the Taiping and Nien Rebellions – that the possibility that their war-fighting effects are on balance positive at least under certain circumstances cannot be ruled out. The question of how these war-fighting processes scale up to affect macro-level war outcomes and postwar regime survival remains central to future research.

Conclusion

Blocking detachments highlight the often-neglected role that coercion can play in motivating soldiers to fight and maintain discipline even under brutal circumstances. The trade-offs and costs associated with

[85] *Washington Post* (2014).

purchasing battlefield staying power are high, however. These costs range from sharply higher casualties to rigid tactics and constrained operations as well as the creation of grievances among soldiers and a second front between officers and soldiers. Though often relegated to the margins of our theoretical and historical accounts of warfare, the role of blocking detachments, as well as similar institutions such as disciplinary units and state-created paramilitaries, opens new avenues for inquiry into how coercion influences military effectiveness.

Understanding these trade-offs will require substantial investment in the collection of microlevel data fine-grained enough to parse out wartime dynamics between blocking detachments, coerced soldiers, local populations, and enemy forces. Ideally, our research design would capture situations where only some units were assigned blocking detachments while other similar ones were not, with selection criteria approaching "as-if" random requirements. Time series data will also become crucial for identifying the effects of these coercive institutions before and after their implementation for a host of battlefield activities, including desertion, defection, and loss–exchange ratios. Part of this inquiry should be directed toward identifying which types of combatants adopt blocking detachments and when they make their initial appearance on the battlefield. There is a key role for qualitative evidence and process tracing in these accounts, too, for many of the measures of battlefield performance, including missed opportunities, decreased flexibility, and poor training, are difficult to gauge quantitatively. Similarly, teasing out the relationship between coercion and desertion will require careful tracing of the sequence of events linking the incidence of soldier indiscipline and the adoption of blocking detachments.

It also remains an open question whether states can "fine-tune" the severity of these trade-offs. It may be possible, for example, to limit the number of executions, or to deploy blocking detachments to only the worst offenders, without undermining the deterrent value of these forces. States could reduce the combat skills necessary for combined arms by adopting less complicated, but more reliable, equipment, thus keeping the force generation properties of coercion while limiting its downside. Where the exact location of the trade-off lies will vary across combatants; it should not be assumed that these trade-offs are necessarily fixed or that states are helpless captives before them. There may be some room to adjust the severity of these trade-offs, if only on the margins, for at least some portion of combatants wielding these forces in battle.

The role of coercion in motivating soldiers, and of blocking detachments more specifically, could also be extended to the study of insurgent organizations. Here, too, coercion has largely been neglected in favor of arguments about the relative importance of material incentives, emotional appeals to revenge, and ideological commitment for recruitment and socialization dynamics.[86] Yet many insurgencies draw on coercive tactics to recruit, motivate, and foster greater discipline. And while insurgent organizations typically favor hit-and-run strikes rather than direct battle, as they edge closer to conventional warfare the formation of disciplinary (sub)units becomes more likely.

These trade-offs may also be altered in the future by the introduction of new technology. While innovations such as drones have clearly affected the lethality of modern warfare,[87] these technologies may have an even greater effect on the ability of belligerents to monitor and sanction their own soldiers. Technologies such as aerostats, drones, and biometric identification could all be deployed to mitigate or even eliminate opportunities for desertion and defection. If blocking detachments were aided, or even replaced, by these technologies, their deterrent effects would be bolstered by removing discretion from local commanders and by closing spaces for soldiers to escape detection. Of course, these C2 systems would also make tempting targets for opposition forces, either via direct action or electronic means that degrade or destroy these battlefield capabilities.

Perhaps most importantly, blocking detachments and other coercive institutions illustrate the importance of casting military effectiveness as the culmination of multiple trade-offs. Militaries are rarely proficient across all facets of effectiveness; choices are often necessary about when to maximize performance in some facets of battlefield performance while accepting greater risk and inefficiencies in others. Efforts to maximize performance in certain areas – say, the reduction of threats to cohesion – can have negative, sometimes unanticipated consequences for other aspects of military performance such as loss–exchange ratios. Studying these trade-offs, including why and when they occur, will push our theories away from simple building

86 Lichbach (1998); Petersen (2002); Wood (2003); Weinstein (2007).

87 See Chapter 6 of this volume.

block approaches and toward richer, more nuanced accounts of military effectiveness.

References

Bartov, Omer. 2001. *The Eastern Front, 1941–45, German Troops and the Barbarisation of Warfare*. 2nd ed. New York: Palgrave.

Beşikçi, Mehmet. 2012. *The Ottoman Mobilization of Manpower in the First World War: Between Volunteerism and Resistance*. Boston: Leiden.

Biddle, Stephen. 2004. *Military Power: Explaining Victory and Defeat in Modern Battle*. Princeton, NJ: Princeton University Press.

2007. Explaining Military Outcomes. In Risa Brooks and Elizabeth Stanley, eds., *Creating Military Power: The Sources of Military Effectiveness*, 207–227. Stanford, CA: Stanford University Press.

Brandenburger, David. 2002. *National Bolshevism: Stalinist Mass Culture and the Formation of Modern Russian National Identity, 1931–1956*. Cambridge, MA: Harvard University Press.

Castillo, Jasen. 2014. *Endurance and War: The National Sources of Military Cohesion*. Palo Alto, CA: Stanford University Press.

Citino, Robert. 2007. *Death of the Wehrmacht: The German Campaigns of 1942*. Lawrence: University Press of Kansas.

2012. *The Wehrmacht Retreats: Fighting a Lost War, 1943*. Lawrence: University of Kansas Press.

Collins, Randall. 2008. *Violence: A Micro-Sociological Theory*. Princeton, NJ: Princeton University Press.

Daines, Vladimir. 2008. *Shtrafbaty i zagradotryady Krasnoi Armii*. Moskva: Yauza.

Ellis, Frank. 2011. *The Damned and the Dead: The Eastern Front through the Eyes of Soviet and Russian Novelists*. Lawrence: University of Kansas Press.

Erickson, John. 1999a. *The Road to Stalingrad: Stalin's War with Germany*. Vol. 1. New Haven, CT: Yale University Press.

1999b. *The Road to Berlin: Stalin's War with Germany*. Vol. 2. New Haven, CT: Yale University Press.

Financial Times. 2014. ISIS Morale Falls as Momentum Slows and Casualties Mount. December 19.

George, Alexander, and Andrew Bennett. 2005. *Case Studies and Theory Development in the Social Sciences*. Cambridge, MA: MIT Press.

Glantz, David. 2005a. *Colossus Reborn: The Red Army at War, 1941–1943*. Lawrence: University of Kansas Press.

2005b. *Slaughterhouse: The Handbook of the Eastern Front*. Bedford, PA: Aberjona Press.

Glantz, David, and Jonathan House. 1999. *The Battle of Kursk*. Lawrence: University of Kansas Press.

2014. *The Stalingrad Trilogy*: Vol. 3, *Endgame at Stalingrad*. Lawrence: University of Kansas Press.

Grossman, Dave. 1996. *On Killing: The Psychological Cost of Learning to Kill in War and Society*. San Francisco, CA: Back Bay Books.

Hamner, Christopher. 2011. *Enduring Battle: American Soldiers in Three Wars, 1776–1945*. Lawrence: Kansas University Press.

Henderson, William. 1985. *Cohesion: The Human Element in Combat*. Washington, DC: National Defense University Press.

Independent. 2015. Suicide Bombers "Defecting from ISIS" and Fleeing to Turkey or Rival Militant Groups. May 15.

International Business Times. 2014. ISIS Executes 100 Foreign Fighters for Trying to Flee Syria. December 20.

Jaggers, Keith, and Ted Robert Gurr. 2004. *Polity IV: Regime Change and Political Authority, 1800–2004*. Ann Arbor, MI: Inter-University Consortium for Political and Social Research.

Khlevniuk, Oleg. 2015. *Stalin: New Biography of a Dictator*. Trans. Nora Seligman Favorov. New Haven, CT: Yale University Press.

Krivosheev, G. F. 2009. *Velikaya Otechestvennaya bez grifa cekretnosti. Kniga poter.'* Moskva: Veche.

Levi, Margaret. 1997. *Consent, Dissent, and Patriotism*. Cambridge: Cambridge University Press.

Lichbach, Mark. 1998. *The Rebel's Dilemma*. Ann Arbor: University of Michigan Press.

Lyall, Jason. 2015. Process-Tracing, Causal Inference, and Civil War. In Andrew Bennett and Jeffrey Checkel, eds., *Process Tracing in the Social Sciences: From Metaphor to Analytic Tool*. Cambridge: Cambridge University Press.

2016a. Paths of Ruin: Explaining Battlefield Performance in Conventional War. Unpublished manuscript.

2016b. Why Armies Break: Explaining Mass Desertion in Conventional War. Unpublished manuscript.

Lyall, Jason, and Isaiah Wilson. 2009. Rage against the Machines: Explaining Outcomes in Counterinsurgency War. *International Organization* 63: 67–106.

Lynn, John. 1984. *The Bayonets of the Republic: Motivation and Tactics in the Army of Revolutionary France*. Urbana: University of Illinois Press.

Marshall, S. L. A. 1947. *Men against Fire: The Problem of Battle Command in Future War*. New York: William Morrow.

Maslov, A. A. 1996. How Were Soviet Blocking Detachments Deployed? *Journal of Slavic Military Studies* 9(2): 427–435.

Merridale, Catherine. 2006. Culture, Ideology, and Combat in the Red Army, 1939–45. *Journal of Contemporary History* 41(2): 305–324.

Millett, Allan, and Williamson Murray. 1988. *Military Effectiveness*. Boston: Allen and Unwin.

Morrow, James. 2014. *Order within Anarchy: The Laws of War as an International Institution*. Cambridge: Cambridge University Press.

Murray, Williamson, and Kevin Woods. 2014. *The Iran-Iraq War: A Military and Strategic History*. Cambridge: Cambridge University Press.

New York Times. 2015. Battered but Unbowed, ISIS Is Still on Offensive. March 13.

Orlov, Andrei. 2012. *Shtrafbat: Prikazano unichtozhit*. Moskva: Astrel.'

Petersen, Roger. 2002. *Understanding Ethnic Violence: Fear, Hatred, and Resentment in Twentieth-Century Eastern Europe*. New York: Cambridge University Press.

Pilster, Ulrich, and Tobias Böhmelt. 2011. Coup-Proofing and Military Effectiveness in Interstate Wars, 1967–99. *Conflict Management and Peace Science* 28: 331–350.

2012. Do Democracies Engage in Less Coup-Proofing? On the Relationship between Regime Type and Civil-Military Relations. *Foreign Policy Analysis* 8: 1–17.

Pogonii, Y. F., ed. 2000. *Stalingradskaya epopeya: Materialy NKVD SSSR i voennoi tsenzury iz Tsentral'nogo arkhiva FSB RF*. Moskva: Zvonnitsa-MG.

Posen, Barry. 1993. Nationalism, the Mass Army, and Military Power. *International Security* 18: 80–124.

Pyltsyn, Aleksandr. 2007. *Pravda o shtrafbatakh: Kak ofitserskii shtrafbat doshel do Berlina*. Moskva: Eksmo.

Quinlivan, J. T. 1999. Coup-Proofing: Its Practice and Consequence in the Middle East. *International Security* 24: 131–165.

Reese, Roger. 2011. *Why Stalin's Soldiers Fought: The Red Army's Military Effectiveness in World War II*. Lawrence: University Press of Kansas.

Reiter, Dan. 2007. Nationalism and Military Effectiveness: Post-Meiji Japan. In Risa Brooks and Elizabeth Stanley, eds., *Creating Military Power: The Sources of Military Effectiveness*, 27–54. Stanford, CA: Stanford University Press.

Reiter, Dan, and Allan Stam. 2002. *Democracies at War*. Princeton, NJ: Princeton University Press.

Reuters. 2015. Islamic State Has Killed Over 2,000 off Battlefield since June: Monitor. April 28.

Rosen, Stephen. 2005. *War and Human Nature*. Princeton, NJ: Princeton University Press.

Rosenbaum, Paul. 2010. *Design of Observational Studies*. New York: Springer.

Rubin, Donald. 2006. *Matched Sampling for Causal Effects*. Cambridge: Cambridge University Press.

Shils, Edward, and Morris Janowitz. 1948. Cohesion and Disintegration in the Wehrmacht in World War II. *Public Opinion Quarterly* 12: 280–315.

Slatta, Richard. 2003. *Simon Bolivar's Quest for Glory*. College Station: Texas A&M University Press.

Snyder, Timothy. 2010. *Bloodlands: Europe between Hitler and Stalin*. New York: Basic Books.

Starikov, Nikolai. 2014. *Voiska NKVD na fronte i v tylu*. Moskva: Algoritm.

Statiev, Alex. 2010. Penal Units in the Red Army. *Europe-Asia Studies* 62(5): 721–747.

Statistical and Accounting Branch Office of the Adjutant General. 1946. *Army Battle Casualties and Nonbattle Deaths in World War II*. Department of the Army.

Stewart, Nora Kinzer. 1991. *Mates and Muchachos: Unit Cohesion in the Falklands/Malvinas War*. Washington, DC: Brassey's.

Stouffer, Samuel, Edward Suchman, Leland DeVinney, Shirley Star, and Robin Williams Jr. 1949. *The American Soldier: Adjustment during Army Life*. Princeton, NJ: Princeton University Press.

Strachan, Hew. 1997. The Soldier's Experience in Two World Wars: Some Historiographical Comparisons. In Paul Addison and Angus Calder, eds., *Time to Kill: The Soldier's Experience of War in the West, 1939–1945*, 369–378. London: Pimlico.

2006. Training, Morale, and Modern War. *Journal of Contemporary History* 41: 211–227.

Sutyagin, Igor. 2015. Russian Forces in Ukraine. RUSI Briefing Paper, March.

Talmadge, Caitlin. 2015. *The Dictator's Army: Battlefield Effectiveness in Authoritarian Regimes*. Ithaca, NY: Cornell University Press.

Telitsyn, Vadim. 2010. *Mify o Shtrafbakakh*. Moskva: Eksmo.

Washington Post. 2014. Desperate for Soldiers, Assad's Government Imposes Harsh Recruitment Measure. December 28.

Weinstein, Jeremy. 2007. *Inside Rebellion: The Politics of Insurgent Violence*. Cambridge: Cambridge University Press.

Weitz, Mark. 2005. *More Damning Than Slaughter: Desertion in the Confederate Army*. Lincoln: University of Nebraska Press.

Wood, Elizabeth. 2003. *Insurgent Collective Action and Civil War in El Salvador*. Cambridge: Cambridge University Press.

Zolotarev, V. A., ed. 1996. *Glavnye politicheskie organy vooruzhennykh sil SSSR v Velikoi Otechestvennoi voine 1941–1945*. Vol. 17 of *Russkii Arkhiv: Velikaya Otechestvennaya*. N.p.: TERR.

5 *Sources of Military Effectiveness in Counterinsurgency*

Evidence from the Philippines

JOSEPH FELTER

What explains why some military forces are effective at combating insurgent threats and others are not? In the context of interstate wars, realist theories posit that a state's capacity to raise and employ military forces able to carry out conventional operations against external threats is well predicted by aggregate wealth and resources. Under these conditions, selection effects within the anarchic international system will theoretically drive states to field the best military forces possible, given their available resources, in order to survive and prevail relative to their peers.[1] A rich literature and multiple comprehensive cross-national data sets exist to inform our understanding of military effectiveness in these conventional interstate conflicts and to assist scholars in making continued advancements in this field.[2]

The effectiveness of forces employed to defeat and deter insurgent threats, however, is more difficult to predict and relies on far less developed theories and empirical data to do so.[3] Allan Millett, cited in the introductory chapter of this volume, defines military effectiveness as "the process by which armed forces convert resources in to fighting power."[4] This implies that greater resources available to states predict increased capacity to fight effectively, which is consistent with realist theories. But realist assessments of a state's conventional military capabilities focusing on material advantages do not necessarily translate

[1] See Waltz (1979).

[2] Highlights include Millett et al. (1988); Biddle (2004); Brooks (2007).

[3] The theoretical literature on counterinsurgency effectiveness draws significantly from the tradition of military science and less from political science. See Galula (1964); Trinquier (1964); Kitson (1974); Thompson (1966); Valeriano and Bohannan (1962). Important exceptions to this military science–dominated literature include Kalyvas (2006); Lyall and Wilson (2009); Lyall (2010); Berman et al. (2011); MacDonald (2013).

[4] Millett et al. (1988, 2).

into explaining its expected capacity to address the complex nature of threats posed by insurgency, where the enemy is often difficult to identify or discriminately target and operates, recruits, and competes with the state from within its own population. Scholars are well equipped to predict and explain military effectiveness in conventional conflict cases such as the decisive US military victory in Iraq following its "hundred-hour war" in 1991 during Desert Storm, for example, but far less capable of similar analyses of the protracted counterinsurgency campaign in Iraq these same US military forces became mired in the following decade.

The primary aim of this chapter is to identify and provide empirical evidence of the sources of military effectiveness in counterinsurgency at the most basic level – the qualities and characteristics of the individual government combatants and the small units that they are assigned. Second, in light of these sources of effectiveness and the resources needed to develop them, I highlight the most salient trade-offs states must make when determining the optimal military forces to field for counterinsurgency based on the nature of the threats they expect to address and important characteristics of their respective domestic political institutions.

I argue that amid the varied and complex environments in which insurgencies operate, the effectiveness of state military forces employed to combat them depends largely on the intrinsic qualities of the individuals and small units conducting these complex operations. The testable implication of this argument is that certain types and configurations of military forces perform better and more efficiently on key metrics of counterinsurgency effectiveness across a range of local conditions under which they are employed.

Anecdotal accounts tying military effectiveness in counterinsurgency to individual- or unit-level qualities are ubiquitous.[5] Empirically rigorous testing of such explanations is scarce in the literature, however, as it requires detailed information on small military units at the tactical level and data on the relative performance of these units in operations under varying threat conditions while holding other potential explanatory factors constant.

[5] Twenty-first-century conflicts in Iraq and Afghanistan, for example, have spawned many notable new contributions to this qualitative literature. See Fick (2006); Mullaney (2010); Kilcullen (2009); Nagl (2014).

To address this limitation, I draw on an original microlevel data set compiled from detailed descriptions of over twenty-one thousand individual conflict-related incidents reported by the Armed Forces of the Philippines (AFP) during government counterinsurgency operations conducted in their country from 2001 to 2008.[6] The data are unprecedentedly rich; they capture wide variation in the qualities of military forces deployed to conduct counterinsurgency, the characteristics of the rebels they confront, as well as the intensity of the threat environment and degree of government control where they operate.[7] Significantly, country level variables shown in previous studies to impact on military effectiveness in conventional settings can largely be held constant.[8]

My empirical analysis of qualitatively distinct military forces under varying environmental and threat conditions demonstrates that well-led, highly trained small units using relevant doctrine are more likely to conduct effective counterinsurgency consistent with the objectives and goals of the civilian leadership they serve. My findings demonstrate that variation in the qualities of the military forces tasked with combating insurgent threats and how they are employed affect important conflict outcomes. Importantly, they indicate this variation is not fully determined by factors such as state wealth or level of development and have important policy implications for the professional training of militaries in reducing the damage from, and possible prospects for, protracted insurgencies and costly civil wars.

Significantly, raising and maintaining military forces capable of effectively addressing insurgency obliges states to make important trade-offs across all three categories this volume identifies: political support, security threats, and war fighting. A number of these trade-offs are distinct from those required when confronting more conventional and symmetric interstate military threats. The types of investments that produce military units effective at combating guerrilla forces competing for support of the local population, for example, are different than those needed to raise and sustain a military capable

[6] See Empirical Studies of Conflict Philippines and https://esoc.princeton.edu/country/philippines. A detailed description of these data is in Felter (2005); Berman et al. (forthcoming).

[7] These data were compiled in direct coordination with senior officials from the Armed Forces of the Philippines beginning in 2003.

[8] Highlights of these factors include culture (Rosen 1995; Posen 1984; Kier 1995; Pollack 2002); regime type/polity (Reiter and Stam 2002); and patterns of civil–military relations (Biddle and Zirkle 1997).

of defending the state against external aggression from a well-armed conventional threat or projecting military power abroad. Balancing a state's interest in fielding forces effective at combating insurgent threats and maintaining internal security with its interests in regime survival – avoiding a coup by this powerful military deployed within its borders – is another example of the trade-offs states must make when determining the quality and make up of its military forces.

This chapter proceeds in five parts. First, I present a bounded definition of what I believe should be the military's expected role when combating insurgent threats in support of the goals and objectives of its civilian leadership. I point to a number of measurable outcomes at the tactical and operational level that aggregate up to support our broad definition of military effectiveness developed in this volume – *the degree to which militaries can accomplish at acceptable costs the goals assigned to them by political leaders*.[9] Second, I identify a select set of individual- and small unit–level characteristics of military forces that I argue are the building blocks predicting military effectiveness in counterinsurgency. Third, I introduce the history of insurgency and counterinsurgency in the Philippines as an especially instructive and generalizable case and describe the original micro-conflict data set used in my subsequent analysis. Fourth, I present results of empirical tests that demonstrate the impact that variation in the qualities of a state's military forces can have on important measurable outcomes when combating a range of distinct insurgent threats. Lastly, I discuss the trade-offs states must make when making investments in their military forces and how my empirical findings can inform resource allocation, force employment decisions and other trade-offs made by states confronting a range of threats posed by insurgency, civil war, and other forms of internal conflict.

Assessing Military Effectiveness in Counterinsurgency

Accomplishing the goals of a state's political leadership is a definition of military effectiveness that can be applied broadly across the spectrum of threats a state confronts. But there can be stark contrasts between employing military force effectively against symmetric foreign-based threats versus those that originate from within the state.

[9] Our definition is laid out in Chapter 1 of this volume.

In conventional interstate warfare, military effectiveness is often manifested in the ability to generate superior relative combat power at decisive times and places on the battlefield in order to destroy an adversary's military forces in the field or seize control of key terrain. Such operations can involve large formations and employ significant firepower. Penalties for collateral damage and incentives for gaining popular support are comparatively limited. Under these conditions, material advantages of states are more readily manifested in the measurable effectiveness of their militaries.

In the small unit guerrilla warfare that characterizes much of insurgent conflict, by contrast, state and insurgent success both turn largely on gaining control and, when possible, the support of the civilian population.[10] Identification of the enemy can be challenging under these conditions and the adverse effects of civilian casualties and other forms of collateral damage particularly severe.[11] Critical objectives of a state's security forces in these types of conflicts are to (1) identify, separate, and discriminately remove insurgents from the population; (2) convince the public that it is safe to support the government – as well as very costly to oppose it – and (3) ensure the popular perception of eventual state victory and insurgent defeat.

Carrots without sticks are generally not effective in cases of suppressing rebellion against the government.[12] While the provision of economic aid and other goods and services as part of targeted efforts to "win hearts and minds" can in some circumstances help increase cooperation and popular support in the near term,[13] they are rarely able to target many of the specific sources and belligerent groups that can pose the greatest risks of rebellion. The credible threat of being captured or killed by government forces is a powerful disincentive for supporting rebellion.[14] In practice, the best-intended efforts of the state to increase livelihood opportunities and raise the living standards of the local populace can be trumped by rebel coercion and intimidation and doomed to fail if implemented prematurely in an environment of insecurity and instability. A modicum of security and credible ability to protect the population is needed to facilitate any government effort

[10] A key finding by Kalyvas (2006) is that cooperation and collaboration follow control in civil wars.
[11] Condra and Shapiro (2012); Berman et al. (2011).
[12] Leites and Wolf (1970). [13] Berman et al. (2011). [14] Gates (2002).

to gain popular support. As Roger Trinquier, an experienced French counterinsurgency practitioner, admonishes, "A few brutalities, such as savagely executed preventive assassinations in the surrounding villages, will cow the inhabitants into providing for the maintenance of the bands and will discourage them from giving useful information to the authorities."[15]

In the many cases of insurgency where the population lives under more Hobbesian conditions, survival will trump moral and principled appeal. Promises by a well-meaning government, or its foreign supporters intent on protecting the civilian population and improving conditions, can be overshadowed by credible and more immediate threats posed by the rebels or a perception that they will ultimately prevail over the government.[16] Given this, military forces employed to combat insurgency must be able to mitigate internal security threats through proactive and discriminate application of force and other actions aimed at the attrition of active rebels and the deterrence of others on the margins from becoming active. In some cases, the optimal application of violence proves enough to defeat the insurgency and consolidate state control while in many others it is a necessary albeit not sufficient prerequisite for government success.[17]

Ultimately, military organizations cannot be expected to unilaterally accomplish the broad goals and objectives of their civilian leaders in a counterinsurgency campaign. Such ambitious objectives, at least in the long term, can only be achieved by the host nation government itself and generally exceed the capacities of state security forces and certainly those of any foreign forces deployed in support of the host nation government's cause. Indeed, state campaigns to combat insurgency have been and continue to be largely unsuccessful in achieving their ultimate political objectives leading many scholars to question the viability of counterinsurgency in general.[18]

While the military alone cannot "win" a counterinsurgency, it can lose it. A military configured and employed to address insurgent threats effectively is a necessary albeit not sufficient component of a comprehensive strategy for ultimate victory. Consider the US

[15] Trinquier (1964, 24). [16] Shafer (1988, 104–132).
[17] Kalyvas and Balcells (2010).
[18] See, e.g., Smith and Jones (2015); Gentile (2013).

experience in Vietnam, where many argue that the firepower-intensive and conventional approach to addressing irregular threats set US forces up for failure.[19]

Thus understanding the sources of effectiveness for this critical tool states employ in COIN – the military – and what it should be expected to accomplish is critical. The components of a more appropriately bounded mission for the military in counterinsurgency, and more reasonable standard to judge military effectiveness in counterinsurgency, include gaining control, establishing security, and – ideally – popular support, whether attracted or coerced, in order to foster conditions and create the space needed for the requisite follow-on whole of government efforts to consolidate control and secure the population.

With these expectations, military effectiveness in counterinsurgency is best achieved by forces able to degrade insurgent capacity by killing, capturing, or facilitating the surrender of rebels and confiscating their weapons while maximizing government force protection and minimizing collateral damage to the civilian population. There are many other useful indicators and outcome measures of military effectiveness in COIN; however, I will argue that military units excelling by these discrete measures are more likely to provide the requisite security and gain the control needed to complement and support a more comprehensive government counterinsurgency campaign.

The selection, preparation and fielding of the forces best able to execute effective counterinsurgency is a bottom-up endeavor and one that rewards investments in the human capital of the security forces in contact with the insurgents and population. I argue that high-quality military forces defined in terms of the leadership, training, and doctrine and the autonomy to initiate operations tailored to local threat complexities are the building blocks that explain effectiveness of military and security forces across a wide variety of locales, conditions, and types of insurgency.

Troop Quality and the Building Blocks of Military Effectiveness in COIN

Counterinsurgency is a difficult mission for any military unit to perform. Information is imperfect; local conditions evolve; and, given the

[19] Krepinevich (1988).

primacy of gaining popular support, the military's objectives are generally much more complex than the attrition of rebel combatants in the field. The US Army–Marine Corps Counterinsurgency Manual quotes a US Special Forces officer's assertion that *"counterinsurgency is not just thinking man's warfare – it is the graduate level of war."*[20] What are the characteristics of the military forces that can be expected to perform effectively in COIN – at this "graduate level" of war? While not an exhaustive list, I describe below the more salient attributes of what I will refer to as "high-quality" military forces that are best prepared to conduct the demanding missions associated with counterinsurgency.[21]

Small Unit Leadership

Capable junior officers and noncommissioned officers (NCOs) empowered to make decisions and exploit opportunities at the small unit level are critical to the success of units combating insurgents. Small squads and teams – a dozen soldiers or less – tasked with enforcing order and maintaining control in remote areas are often the only representatives of government that the local population has contact with. Very often it is at this level that hearts and minds are won or lost and the verdict on support for the government is de facto determined. Small unit leadership plays a significant role in bringing the military's capacity to bear against insurgents effectively and discriminately. Leaders at these tactical levels – more intimately than senior commanders or policy makers – ultimately determine whether a state's resources are leveraged effectively to gain and maintain control of the population, provide security, and credibly signal eventual triumph by the forces of the state. The difference between a disciplined military unit able to promote government policies and enforce order more effectively and an unruly band of thugs prone to engage in rape, pillage, and plunder, for example, can be determined by the presence of professional junior officers and NCOs maintaining good order and conduct among the troops.

[20] US Army and US Marine Corps (2007, Chapter 1).

[21] The qualities identified here are informed by the military science literature and the author's experiences over a twenty-five-year career as a US Army Special Forces officer with extensive experience conducting foreign internal defense and security assistance missions and commanding the ISAF Counterinsurgency Advisory and Assistance Team in Afghanistan in 2010–2011.

Training and Doctrine

Tough, realistic training is the hallmark of effective military units and a significant predictor of success in all operational activities. A small military unit will not be effective in the operations and tactics a counterinsurgency environment demands if it is not trained to perform them. Quality training can be expensive to conduct, however, and it varies across and within military organizations. In militaries that rely on conscription, for example, training is challenged by the turnover of personnel and the shortage of career soldiers providing continuity of expertise. Within the same military, some units are provided greater resources for training or get more out of the budgets provided them. Organizational norms within certain military units affect training levels as do the priorities individual commanders place on the planning and execution of tough mission focused training plans incorporating threat appropriate doctrine.

Decentralized Control

Combating insurgency demands agile forces able to anticipate, create, and exploit opportunities at the local level while ensuring their actions are consistent with the overall intent of the government's broader counterinsurgency campaign. Flexibility at the small unit level is critical – well-trained and professionally led small units will be ineffectual if a micro-managing chain of command restricts their freedom of action or if an internally imposed unwillingness to act without clearance from higher authorities prevents a unit from seizing fleeting local opportunities to address insurgent threats. Acting on such opportunities in dynamic threat environments requires agility and decisiveness at the tactical level. The hierarchical nature of many conventionally trained and employed military organizations can be detrimental to these ends. Organizational norms in some units that inhibit subordinate freedom of action exacerbate this effect.[22] An effective counterinsurgency force must be able to enforce centralized intent and monitoring without inhibiting decentralized execution and responsiveness to unique local conditions. These building blocks of individual and small unit–level

[22] Pollack (2002); Biddle and Zirkle (1997).

military effectiveness presented here may strike many practitioners as near tautological. Military units that possess high-quality leadership, training, doctrine, and significant autonomy at the small unit level are more capable of taking advantage of emerging opportunities, proactively conducting operations, and seizing the initiative from insurgents. That investments in the human capital of counterinsurgency forces can develop these attributes and lead to more effective outcomes in COIN is not surprising. Empirical evidence of the degree to which differences in the individual- and small unit–level qualities of military forces conducting counterinsurgency impacts important conflict outcomes and under what conditions, however, is rare in the literature and has significant policy relevance. Isolating variation in the quality of military forces conducting COIN and testing the impact of the building blocks of military effectiveness presented here is challenging. At a minimum it requires a case or set of cases of counterinsurgency with detailed information on the qualities and structures of the state's internal security forces, how they are employed at local levels as well as micro-data on important measurable outcomes resulting from their individual operations and activities.

An Ideal Test Case: Insurgency and Counterinsurgency in the Philippines

The multiple distinct types of insurgency in the Philippines and variation in the quality of forces the government responds to them with make it an ideal case to study and one that can be generalized to inform our understanding of military effectiveness in COIN across a range of other cases. Home to some of the most protracted insurgencies in the world, the Philippines currently has a communist, separatist, and extremist insurgency active in the country. Each of these rebel groups have distinct motives, have recruitment bases, and conduct their struggle against the Philippine government differently, both at the political level and at the tactical level, against its security forces in the field. The Armed Forces of the Philippines (AFP) tasked with combating these distinct insurgent threats employs security forces that differ sharply on the dimensions of quality described above. Key attributes of these insurgent groups and of the distinct types of state military forces employed against them are described below.

The Insurgents: Active Insurgencies in the Philippines

Communists

The Communist Peoples Party (CPP) and its armed wing the New Peoples Army (NPA) have been waging a classic Maoist people's war aimed at overthrowing the Philippine government since 1968. At the height of its strength in 1986, more than twenty-five thousand armed rebels belonged to the NPA. At that time, the NPA's political influence extended to sixty-three of the then seventy-three provinces in the country.[23] Since the mid 1980s, the NPA has declined in strength. However, in the early 2000s, the CPP-NPA enjoyed a resurgence and spiked to over eleven thousand armed rebels in 2001. It remains a significant threat to the government, with over five thousand armed rebels still considered active in 2008.[24] The NPA's survival depends on popular support, whether attracted or coerced. Small unit harassment and hit-and-run tactics characterize NPA military activities. NPA insurgents embrace Mao's concept of Peoples War, fight asymmetrically, and rarely mass against better-armed government soldiers or engage in direct attacks.[25]

Separatists

In the southwest provinces of Mindanao island and the Sulu sea, members of the Moro Islamic Liberation Front (MILF) continue the Bangsa Moro[26] struggle for independence from what they perceive as an illegitimate Christian government in Manila.

The MILF's core grievances stem from the Philippine government's efforts to retitle lands that the MILF consider part of their ancestral domain as well as the historical marginalization and disenfranchisement of the Filipino Muslim population. Because of the perceived

[23] Estimates provided by the Philippine Army Deputy Chief of Staff for Intelligence (G2).

[24] These figures were provided by the Armed Forces of the Philippines General Headquarters (AFP GHQ).

[25] Mao (1965, 124) describes the principal tactics of guerrilla war: "The enemy advances, we retreat; the enemy camps, we harass; the enemy tires, we attack; the enemy retreats, we pursue."

[26] Bangsa Moro refers to Muslims in the Philippines who have resisted Christian domination for centuries.

legitimacy of the MILF's agenda among much of the Islamic population, the MILF enjoys broad active and passive support from Muslim communities in which they operate. The MILF claim specific geographic areas, and unlike communist insurgents, MILF units are known to mass forces and fight government units in semiconventional military engagements that include pitched battles. The MILF strength peaked at approximately sixteen thousand armed rebels in 1999 before suffering significant attrition as a result of a major offensive by the AFP in 2000. In 2008, the latest year included in this study, the MILF was believed to have close to twelve thousand members.

Extremists

The Abu Sayyef Group (ASG) operates mainly in the southern islands of Basilan and Sulu Sea and to a limited degree within the southwest portions of the island of Mindanao. The group claims to maintain a separatist agenda but engages in extensive criminal activities and terrorism. The ASG is known to have linkages with international terrorist groups, including al-Qaeda, and has been active in kidnap-for-ransom and other criminal and terrorist activities since the mid 1990s.[27] The ASG's motives are largely based in criminality, and thus the group is widely acknowledged to be less concerned with gaining and maintaining popular support than the communist NPA or separatist MILF. Using resources gained from illicit activities and its alleged support for international terrorist groups, the ASG is well armed and equipped and able to readily intimidate the local population. The ASG operates in mobile guerrilla bands and conducts ambushes and harassments of government forces and has been responsible for the deliberate killings of civilians on many occasions. The ASG swelled to over one thousand members at the peak of an international hostage crisis it instigated in 2000 and has subsequently suffered drastic attrition and was believed to be under four hundred members by 2008. This precipitous decline was aided by significant increases in US military assistance and advisory support as part of the global response to terrorism following the 9/11 attacks.

[27] Abdurajak Janjalani, a founding member of the ASG, fought in Afghanistan with the Mujahedeen in the 1980s. More recently, in 2017 Abu Sayyef Group senior leader Isnilion Hapilon was declared "Emir" of the Islamic State of Iraq and Syria (ISIS) in Southeast Asia by ISIS leader Abu Bak al-Baghdadi.

Table 5.1 *Select Characteristics of Philippine Insurgent Groups*

	Communists (NPA)	Separatists (MILF)	Extremists (ASG)
Objectives	Replace government	Independence	Mixed
Popular support	Contested	Ethnic/tribal	Ethnic/tribal/ purchased
Strategy	Population focused	Terrain focused	Terrorism/ predation
Tactics	Guerrilla	Semiconventional	Guerrilla/ terrorist
Recruiting	Poverty	Religion	Religion/ criminality

A comparison of the three main insurgent groups active in the Philippines today along key dimensions is provided in Table 5.1.

Philippine military forces are employed against this wide variety of rebel threats operating under the same or similar terrain and other environmental factors. Assessing the effectiveness of these forces across these distinct threat groups increases the degree to which the results can be generalized across other cases in countries where government forces confront revolutionaries, separatists, and extremist threats. The characteristics of the Philippine military forces conducting counterinsurgency vary sharply as well and are described in the next section.

The Counterinsurgents: The Armed Forces of the Philippines

The Armed Forces of the Philippines (AFP) consists of approximately 126,000 soldiers, sailors, airmen, and marines and has the lead mission for providing internal security and combating the multiple insurgent threats active in the Philippines.[28]

Augmenting these active duty forces are over fifty thousand locally recruited and employed part-time auxiliaries. These forces vary in terms of quality and ability to gain access to local information and intelligence. The Philippine military can be divided into three broad qualitative categories – regular forces, Special Operations Forces (SOF), and local forces – that differ along these and other dimensions.

[28] These figures are accurate for the last year of the data studied.

A general description of these three types of forces along with their relative strengths and weaknesses is provided below.

Regular Forces

Primary responsibility for internal security operations in the Philippines falls to the eight thousand strong Philippine Army organized around ten regular infantry divisions with three infantry brigades assigned to each as well as three marine infantry brigades from the Philippine Marine Corps. While internal security is the primary mission of the Philippine military, much of the formal training and doctrine for its regular forces is geared toward preparing to wage conventional operations against external threats.

Special Operations Forces (SOF)

Special Operations Forces (SOF) units around the world are known for their capable junior leaders, intense training, discipline, and an organizational culture that promotes innovative problem solving and small unit initiative. SOF units from the Armed Forces of the Philippine share these same characteristics and attributes relative to other units in their military. AFP SOF units include the First Scout Ranger Regiment and the Special Forces Regiment of the Philippine Army Special Operations Command. Also included in this category are units assigned to the Philippine Navy's Special Warfare Operations Group and the Philippine Air Force's Special Operations Wing.

Local Forces

Providing security and holding areas at risk of rebel infiltration exceeds the capacity of the AFP's active forces and requires augmentation by part-time security forces. The Citizen Armed Forces Geographical Units Armed Auxiliaries or CAAs are the bulk of this force and are recruited and deployed locally to combat insurgency and other threats to internal security. In 2004, the middle year of the sample studied, the Armed Forces of the Philippines records accounted for 3,246 detachments of CAAs with a total of 52,748 personnel deployed around the Philippines.[29]

[29] These data were compiled from records maintained at the General Headquarters, Armed Forces of the Philippines.

The AFP's SOF, regular forces, and local forces differ sharply in terms of quality. A relative assessment and comparison of these military forces along the key attributes identified as critical to counterinsurgency effectiveness is provided below.

Comparing Troop Quality

Leadership

SOF units enjoy superior leadership relative to the AFP's regular and local forces down to the small unit level. For example, the Philippine Army's elite Scout Rangers and Special Forces regiments have on average 100 percent of the authorized number of officers at the company level (five) compared to regular infantry battalions who are assigned on average only two – less than half the doctrinally authorized numbers of officers with many infantry companies being led by Lieutenants versus Captains.[30] The typical Company Commander in the Special Forces or Scout Rangers, in contrast is a Captain that has served in leadership positions for five years or more with formal mentoring opportunities to help develop their capabilities and prepare them for command. Also significant, the quality of NCOs in SOF units is generally far superior to that of those serving in regular infantry units. SOF NCOs are provided much greater formal professional development opportunities as well as experience in assuming leadership positions in the small teams that are the operational elements of these forces.[31] On the other extreme, local forces are dependent on the leadership provided by cadres from the regular infantry or special forces units.

Training and Doctrine

SOF units in the AFP receive considerably more and higher-quality training than their regular and local force counterparts. Scout Rangers and Special Forces members each undergo a grueling seven-month qualification course over and above the initial training received by regular infantry units. These courses are geared toward preparing them

[30] These statistics provided by the Armed Forces of the Philippines Deputy Chief of Staff for Personnel (J1).

[31] This qualitative distinction is perhaps more significant, given the emphasis COIN places on small unit operations.

Table 5.2 *Comparing Building Blocks of Effective COIN Forces*

Unit	Leadership	Training and doctrine	Decentralized control	COIN force quality
SOF	High	High	High	High
Regular	Medium	Medium	Medium	Medium
Local	Low	Low	Low	Low

to operate in small teams and engage in the guerrilla warfare required to interdict many of the insurgent threats they will face while regular forces train and follow doctrine aimed at addressing more conventional threats. These courses and multiple additional specialty courses provided to SOF personnel far exceed the three-month basic training provided regular forces and the three weeks of formal military training local forces receive.

Decentralized Command and Control

SOF units operate in small teams and deploy much more independently than their regular infantry counterparts. The culture in Philippine Army SOCOM is to empower NCOs and give them greater responsibility to plan and execute operations whereas in regular infantry units, decision-making authority is much more centralized and maintained almost exclusively by commissioned officers. Local forces are widely dispersed in remote areas far from the supervision of senior leadership but typically not empowered to initiate operations at their level.

Table 5.2 describes and summarizes the comparisons of relative qualities of SOF, regular, and local forces across the key dimension described above.[32]

Data Sources and Methods

When small units of SOF, regular as well as local forces are deployed under similar environmental and operational conditions, much can be

[32] This ordering is based on a review of official records of accession standards, quality and professionalism of leadership, degree and intensity of training, and the appropriateness of doctrine to fighting insurgent threats.

learned about how the differences in their characteristics and employment influence their expected performance and effectiveness. These tests and comparisons are made here with data comprising details of over twenty-one thousand insurgency and counterinsurgency incidents involving the Armed Forces of the Philippines from 2001 to 2008.[33] These data are compiled from the original field reports of every operational incident reported to the Armed Forces of the Philippines's Joint Operations Center by deployed military units conducting internal security operations against ongoing insurgent and terrorist threats around the country. Highlights of the information coded from these incident level reports include the date, location, participating government units, type of rebel group involved and a description of each incident along with measurable results in terms of government, rebel, and civilian casualties; fire arms recovered/lost; and the number of rebels surrendered, captured, and apprehended as a result of the operation.[34] During the time period studied, the Armed Forces of the Philippines reported 2,550 government troops and 2,852 rebels killed as part of operations; 1,808 civilian deaths were recorded, for a total of 7,127 deaths reported as a direct result of operational incidents for the eight years studied.[35]

Testing the Impact of COIN Force Quality on Military Effectiveness in COIN

It is challenging to disaggregate the impact of military activities from the broader government counterinsurgency efforts they support as well

[33] The author is indebted to the Chief of Staff of the Armed Forces of the Philippines and the staff in the Office of the Deputy Chief of Staff for Operations (J3) for providing unprecedented access to these data. The data used for this study are in unclassified form.

[34] Each individual incident is assigned a unique location identification number, which allows it to be plotted down to the Philippine village level using maps that interface with software from the Geospatial Information System (GIS). This enables the data to be presented visually on an integrated map of the Philippines and analysis of the operational incidents along several physical and spatial dimensions of interest.

[35] These figures were compiled from the results of all operational incidents reported by the AFP during that time. Deaths that resulted from nonoperational incidents, such as vehicular accidents, disaster assistance, unexploded ordinance, and other types of incidents, were not included. Additionally, the significant number of civilian casualties that resulted from conflict-related actions and the aftermath of conflict are not accounted for here.

as from other environmental factors unrelated to government interventions. Measuring the performance of military forces in the individual conflict episodes they are involved in, however, provides an opportunity to assess and compare the relative effectiveness and potential of different types and configurations military forces in COIN at its most basic unit of analysis – the operational incident. The analysis that follows measures and compares the performance of comparably sized units conducting similar operations against the same rebel opponents and on similar terrain. Well-led, highly trained, semiautonomous small military units make the most effective counterinsurgents. Higher-quality SOF units inflict on average greater rebel casualties, suffer fewer casualties themselves and fewer civilian deaths in the incidents they are involved in relative to the medium-quality regular infantry and lesser-quality local forces respectfully. Significantly, the superior performance of higher-quality COIN forces is more pronounced when they are employed against asymmetric threats such as the communist New Peoples Army than when confronting separatist rebels of the Moro Islamic Liberation Front who often wear uniforms, hold territory, and are known to fight using semiconventional tactics. The mean reported size of government forces involved in the reported operations in this sample is seventeen, with SOF incidents averaging nineteen soldiers, regular units twenty-two, and local forces thirteen.

Table 5.3 posts results from pooling the total number of operational incidents reported from the field by the Armed Forces of the Philippines during the sample period and isolates those where SOF, regular, and local forces participated unilaterally versus combined operations with other forces. The results are broken down further by major insurgent group involved that allows comparisons of outcomes based on distinct types of rebel opponents.[36]

These findings, while somewhat intuitive, provide rare empirical evidence that high-quality military forces conducting counterinsurgency operations are more effective in achieving outcomes associated with military effectiveness in counterinsurgency at the tactical and

[36] In a large number of incidents, various combinations of forces operated or were involved together, and the sum of the incidents where SOF, regular, and local forces operated unilaterally falls well below the total incidents reported for this period. Additionally, several other smaller and/or spin-off groups beyond communist, regular, and local are included in the total; thus the sum of these three main insurgent groups does not equal the total posted.

Table 5.3 *Average Battle Deaths per Incident from Counterinsurgency Operations in the Philippines, 2001–2008*

Unit participant (# unilateral incidents)	Rebels killed	Government killed	Civilians killed
SOF	**0.38**	**0.11**	**0.03**
vs. communists (258)	0.31	0.10	0.01
vs. separatists (61)	0.21	0.15	0.01
vs. extremists (146)	0.67	0.12	0.04
Regular	**0.21**	**0.15**	**0.05**
vs. communists (3649)	0.20	0.13	0.02
vs. separatists (618)	0.28	0.15	0.10
vs. extremists (296)	0.41	0.39	0.01
Local	**0.17**	**0.30**	**0.08**
vs. communists (419)	0.17	0.40	0.05
vs. separatists (220)	0.15	0.15	0.08
vs. extremists (41)	0.34	0.68	0.10

operational levels. The results in Table 5.3 are interpreted further and discussed in greater detail below.

Raising the Costs of Insurgency (Rebels Killed)

When Philippine SOF units meet rebels in combat operations, they inflict considerably higher mean casualties per incident than do regular infantry and local forces.[37] In the sample period studied here, rebels suffered nearly 50 percent greater casualties on average when they met

[37] Heightened US SOF led advisory and assistance support provided to the AFP since early 2002 focused on building capacity and executing operations to interdict the extremist Abu Sayyeff Group. AFP SOF units were given operational priority by US advisers, but all AFP efforts to combat the localized ASG threat were enhanced both by improving the quality of the AFP forces involved through training and advisory efforts and by increasing integration to intelligence provided by US assets and resources. Multiple anecdotal accounts and author experiences suggest that the especially high average rebels killed in operations between SOF and the extremists (ASG) as well as somewhat higher average rebels killed per incident for extremists by regular and local forces are driven in part by the effectiveness of the advisory and assistance support provided by JSOTF-P and other US efforts during this period. The classified nature of these activities and corresponding data restrictions make it difficult to test this hypothesis empirically.

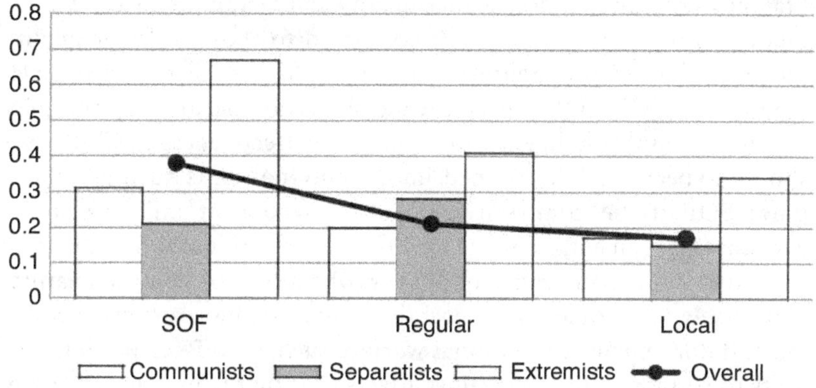

Figure 5.1 Average rebel casualties per operational incident, 2001–2008.

SOF units in individual combat incidents than they did against regular infantry units and over twice as many more than the number of rebel casualties incurred in incidents involving only local forces (Figure 5.1).

The explanation for these outcomes consistent with my theory is that better-led and -trained forces empowered to take initiative at the small unit level can be expected to inflict higher mean casualties on rebels in the combat operations they are involved in. Skeptics of this explanation might argue that the observed differences in mean rebel casualties per combat operation are determined more by where these units are deployed and how they are supported than by the capacities and structures of the units themselves. For example, it may be that SOF forces are sent to "target-rich areas" where the opportunity to engage rebels is greater and that the lower priority assignments are given to regular infantry units and even lower priority areas relegated to the locally recruited forces. These and related conclusions are not supported by the data. When the aggregated incident level performances of forces across these quality categories are compared under varying types of rebel opponents, proximities to rebel presence, and intensity of rebel activities, the higher-quality SOF units inflict greater losses on insurgents under all conditions tested.[38] Whether operating in a

[38] Data used to test this alternate explanation include information on the location of ten known MILF bases and the location of 2,762 villages deemed affected by Communist rebels as of 2004 – the midyear point of this study. Using these data and GIS tools, the operational incidents were compared at varying

"target-rich" area or not, higher-quality military units in the Philippines perform better on average compared to those of lesser quality under similar threat conditions at the individual conflict incident level.

Interestingly, if COIN-specific leadership, training, and doctrine play a critical causal role in explaining military effectiveness in COIN, we should expect specially trained unconventional units such as Philippines SOF to be comparatively more effective at inflicting insurgent casualties in engagements against unconventional threats fighting asymmetrically than against opponents of a more conventional nature. Conversely, we should also expect that units with more extensive training and doctrine in conventional warfare such as AFP regular infantry units to be more effective against adversaries that fight conventionally or semiconventionally.[39] Broken down by insurgent group, the insurgent casualties results are consistent with this intuition: SOF units, on average, kill 80 percent more insurgents per incident than regular infantry units and 123 percent more than local forces when confronting communist and extremists guerrillas but fare comparatively much worse against the separatist MILF, which engages in more semiconventional pitched battles, thus allowing the AFP to employ conventional tactics, such as massing superior firepower, e.g., artillery, against them. The data support this observable implication of my theory as regular units inflict greater casualties on the semiconventional MILF rebels at the incident level for the years studied.[40]

Force Protection and Effective COIN (Government Killed)

In combat operations reported during 2001–2008, Army Special Forces, Scout Rangers, and other SOF units suffered nearly one-third fewer average losses per incident than were incurred by regular infantry

distances to MILF bases as well as by those occurring in a municipality that contained at least one communist affected village versus those free of communist rebel influence. SOF kill more enemy per combat incident under all these varying threat conditions. See Felter (2005) for results of these tests using 2001–2004 incident data.

39 Thayer (1985) shows a similar trend prevailed during the Vietnam War. Elite provincial reconnaissance units inflicted Vietcong casualties much more efficiently than regular units.

40 Multiple author interviews with Philippine military members suggest that fighting the more elusive communist guerrillas requires much more expertise and specialized skill than engaging the comparatively more static and easier to identify separatist rebels, such as those from the MILF.

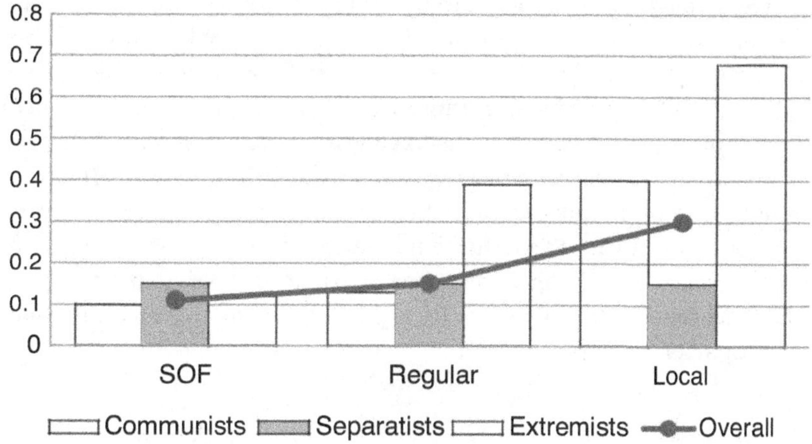

Figure 5.2 Average government casualties per operational incident, 2001–2008.

units. During this same period, local forces suffered twice as many average battle deaths as regular forces and nearly three times more than SOF (Figure 5.2).

According to my theory, these results can be largely attributed to tactics, force protection, and operational security factors best explained by variation in small unit leadership, training, and discipline. Limiting friendly casualties from operations demands vigilance at the small unit level, especially under the field conditions and protracted threats facing the Armed Forces of the Philippines. Enforcing strict security and force protection measures such as aggressive patrolling and sustained security operations requires hands on leadership and close supervision at the squad level. Sound standard operating procedures, immediate action drills, and contingencies for reacting to enemy contact must be reinforced through demanding and realistic training. Local forces arguably are killed at relatively higher rates because of inferior efforts to protect themselves, a comparatively mediocre operations tempo, and poor tactics, all of which stem from poor leadership, discipline, and training.[41]

[41] In fact, one of the highest casualty-producing incidents suffered by the AFP during this eight-year period occurred during a raid by communist insurgents on a militia detachment in the remote village of Cagpili in Eastern Samar on June 26, 2003. An estimated two hundred insurgents overran the detachment

The considerably higher casualties suffered by local forces in combat operations could be attributed to the location where these troops are employed. Local forces are typically deployed in more remote areas far from military bases and support from other units. This challenges the ability of these detachments to call on reinforcements, employ fire support assets, receive timely medical evacuation, and benefit from other types of support from higher headquarters. The data, however, do not support this alternative explanation. At the individual incident level, local forces suffer higher mean casualties in combat operations than do regular forces and SOF, regardless of the distance to higher headquarters.[42]

Discrimination in Conflict Episodes (Civilians Killed)

The AFP reported 1,725 civilian deaths resulting from operational incidents around the country from 2001 to 2008.[43] The mean reported civilian deaths resulting from operational incidents for all three government unit categories during this period, encouragingly, was much lower than the average battle deaths for government and rebel combatants. One explanation for this relatively low number of civilian deaths across different unit types is that the AFP tends not to rely on firepower-intensive operations, however, there remains relatively significant variation in civilian casualties across COIN unit types. My theory predicts that strong leadership from NCOs and junior officers provided to

and killed seventeen militia members and their regular army cadres before making off with twenty-four government firearms. After-action reports on the incident provided to the author by the Philippine Army Assistant Chief of Staff for Operations (G3) blamed the detachment for not rehearsing its actions on contact drills and plans as directed by their detachment commander just days before the rebel assault.

42 The measure for proximity to higher headquarters used to test this was the distance between incidents and the nearest AFP brigade headquarters. While not perfect, it does offer a reasonable proxy for access to greater support, logistics, and reinforcements. In this sample, government units suffered close to the same mean killed in action per incident, whether it occurred within twenty-five miles of one of the twenty-seven brigade headquarters or over seventy-five miles away. SOF and regular forces do not experience a major difference in expected casualties for operations very close or very far from the brigade headquarters either.

43 This figure reflects civilian casualties reported by the AFP as a result of operations and is far short of total civilian deaths attributed to conflict. Estimates of civilian casualties vary widely across sources, and many are much higher than official AFP estimates.

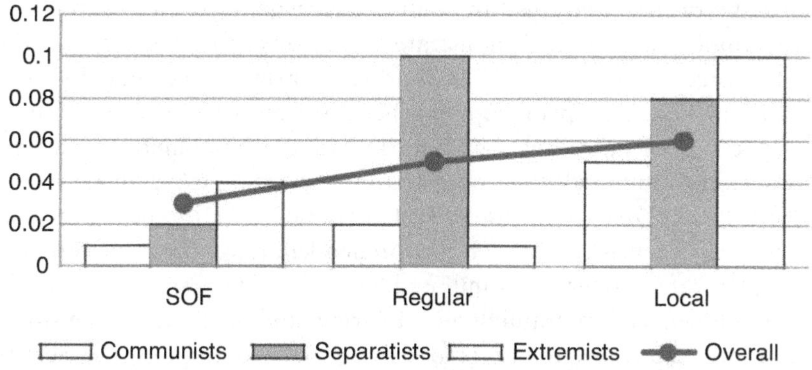

Figure 5.3 Average civilian casualties per operational incident, 2001–2008.

well-trained and disciplined troops, insures more discriminate operations are conducted. Thus we expect that SOF units will inflict fewer civilian casualties than regular and local forces. By virtue of their professionalism, regular units should rank second. With comparatively less training, weaker leadership, and limited professionalism, local forces should rank third.

The evidence supports these intuitions: SOF units used force with greater discrimination and with less harm to the civilian population. The chances of a civilian casualty occurring in an incident between a SOF unit and insurgents is only 0.03, or about one civilian death per thirty-three incidents. The chances of a civilian dying as a result of an engagement between a regular infantry unit and insurgents increases to 0.04, or about one civilian death per twenty-five incidents.[44] When local militias engage insurgents, civilians are twice as likely to be killed compared to SOF-insurgent engagements or about one death per seventeen incidents. Importantly, civilian deaths that occurred as a result of rebel actions without government forces present are often not reported in these data (Figure 5.3).

None of the measures identified above are meant to suggest that battle deaths are an appropriate lone metric to assess the effectiveness

[44] Operations pitting regular forces against separatists from the MILF result in much greater civilians killed than when they confront communists or extremists. Regular soldiers with significant experience fighting the MILF report that women and boys will often support their husbands and fathers in combat bringing ammunition and other supplies thus inserting themselves in more dangerous situations and making it difficult to determine combatant or non-combatant status in the heat of battle. Also, the fact that these rebels have

of military forces engaged in counterinsurgency over time. However, government units that demonstrate the capacity to inflict damage on rebel forces in operations while limiting their own and civilian losses, I argue are higher-caliber troops in general and more likely to excel along a spectrum of other missions both kinetic and nonkinetic in nature. Additional empirical tests indicate that increasing troop quality also predicts other relevant incident level outcomes in COIN such as greater numbers of rebels captured and apprehended, fewer firearms lost, and a higher rate of operations initiated.[45]

Sound leadership, training and doctrine, and flexibility at the small unit level can translate into responsive execution of a range of activities counterinsurgency demands. These high-quality units, if employed and supported effectively by senior military and civilian leadership, will be better able to exert control in the areas in which they operate, thus reducing the local populace's estimated risks and increasing their incentives to cooperate with the government.

Trade-offs in Employing Effective Military Forces for Counterinsurgency

Empirical analyses of the counterinsurgency experiences of qualitatively distinct Philippine military forces in this chapter provide evidence that bottom-up investments in the quality of the individuals and small units employed to combat insurgents increase the effectiveness of these forces and enable them to achieve more favorable tactical and operational level outcomes in conflict episodes. But fielding specially trained forces like the Philippine Army's Scout Rangers and Special Forces and other elite units in militaries around the world is expensive and takes time, and these forces cannot be mass-produced without diluting their quality.

Additionally, the advantages in effectiveness that higher-quality COIN forces such as the Philippine SOF forces analyzed here possess are contingent on the nature of the threats they are employed against. Investments in the human capital of COIN forces have a high

fixed bases where their families and civilians live put them at greater risk when attacked by government forces.

45 See Felter (2005) and Crost et al. (2014) for evidence that increasing troop quality predicts superior performance across a range of incident level outcome COIN outcomes.

return when confronting asymmetric irregular threats like the NPA guerrilla fighters living among the population but less so when confronting semiconventional and conventional rebels or external enemies of the state. In a resource-constrained environment, trade-offs in determining the quality versus quantity and quality versus firepower of COIN forces must be made in light of the nature of the threat the military will be addressing. Lastly, (over)investing in the building blocks needed to field high-quality military forces effective at accomplishing the goals of their political leaders can, in some cases, undermine the interests of these same political leaders in an extreme way – high-quality small units empowered to take the initiative to defeat insurgents are that much more capable of launching a successful coup against the state.

With this appreciation of the demonstrated advantages of fielding high-quality COIN forces and the costs and the constraints associated with this, states must make trade-offs based on their resources available, prevailing security threats, and domestic political considerations. These trade-offs span the broad categories of political support, security threats, and war fighting described in the introductory chapter of this volume, and several of the more salient of these trade-offs in these respective categories are discussed below.

Political Support

The quality of a state's military and how it fights and operates can present trade-offs relating to political support. For example, high-quality COIN forces are better able to limit civilian casualties in the operations they are involved in for reasons I argue stem from heightened levels of discipline, professionalism, fire control, and other factors related to their superior small unit leadership and training. Civilian casualties can undermine popular support for government or third-party forces conducting counterinsurgency. States are confronted with a trade-off between raising and maintaining more expensive, higher-quality COIN forces with increased capabilities or cheaper, lower-quality COIN forces that may be less discriminate or capable of limiting civilian casualties. Additionally, these same high-quality forces are better able to limit the casualties they suffer themselves in counterinsurgency operations, and this can also be a key driver of domestic political support for sustaining a counterinsurgency campaign. This is especially true in third-party counterinsurgency operations such as the US conflicts in Vietnam, Afghanistan, and Iraq.

Thus states involved in, or preparing for, counterinsurgency campaigns where significant penalties exist for incurring civilian casualties – or casualties among their own forces – have stronger incentives to invest in high-quality COIN forces better able to mitigate the risks of incurring these casualties. States involved in COIN campaigns where civilian casualties and collateral damage are deemed less consequential to accomplishing their goals at the strategic or domestic political levels, however, may calculate a different utility for high-quality forces and make compromises and trade-offs accordingly.

An even more challenging trade-off states make when raising forces to combat insurgency stems from the fact that the more capable and effective a state's military is at addressing internal threats to national security, the greater the potential that these capacities can be turned against the established government. Succinctly put, investments to boost the effectiveness of military and security forces for COIN can increase risks of coups. Loyalty and support for the regime are purchased in some cases at the expense of effectiveness.[46] An added concern impacting this trade-off is that the cohesion, identity, and sense of loyalty developed in higher-quality elite military units can help overcome collective action problems and increase their ability to launch a coup. Thus, somewhat ironically, investments in the professionalism, training, and other aspects of the human capital in a state's military can increase its effectiveness in the field by host of measurable outcomes but also increase the risk that these same forces and their capable leaders turn these capacities against the regime of the state.[47]

Clear evidence of this phenomenon is observed in the Philippines in recent decades where some members of the Scout Ranger Regiment – the most effective SOF unit in their military when employed against insurgents – were also involved in multiple coup attempts throughout the late 1980s and early 2000s. In the aftermath of a 2003 destabilization plot led by young SOF officers – many of them recipients of extensive US counterterrorism training – the political leadership and

[46] Pollack (2002); Biddle and Zirkle (1997). States can take steps to "coup-proof" their forces. Carey et al. (2015).

[47] Savage and Caverley (2014) and Reiter (2001) show a relationship between US international military exchange training and risk of future coups.

senior leaders in the Armed Forces of the Philippines opted to drastically reduce the size of the Special Forces and Scout Ranger Regiment as well as populate the chain of command of the remaining SOF units with officers drawn from the regular infantry and other sources outside of the Ranger and Special Forces.[48] This is a stark example of attempts to "coup-proof" the force and of a state trading off military effectiveness for political reliability and loyalty among the ranks.

Security Threats

The principal trade-off states must make when determining the optimal investment in the quality of their COIN forces is driven by the nature of the threats they expect to confront. Raising the types of forces capable of effectively addressing asymmetric insurgent threats, such as guerrillas operating among the population, requires different investments and priorities than for forces employed against conventional interstate and more symmetric insurgent threats.[49] Superior firepower is much more difficult to bring to bear against asymmetric insurgent threats operating among the population where collateral damage imposes costs on government security forces and thus a much less important component of military effectiveness compared with military operations in a symmetric threat environment such as conventional interstate wars.

States who only expect to face asymmetric insurgent threats have a greater incentive to invest and employ the types of high-quality forces described in this chapter. These states' interests are not threatened to a large degree by a lack of investment in expensive weapon systems or other military hardware. States confronting conventional or symmetric semiconventional threats can expect a smaller relative return on investments in high-quality COIN forces and a greater requirement for force size and firepower in their expected operations and campaigns. Thus they may opt to field cheaper, lesser-quality forces and increase expenditures on numbers of soldiers, firepower, and military hardware based on the nature of the threat they anticipate addressing.

[48] General Efren Abu, the Commanding General of the Philippine Army in the aftermath of the 2003 coup attempt, called this reduction of the Special Forces and Scout Rangers "right sizing" the force. Author interview, August 2004, Philippine Army Headquarters, Fort Bonifacio, Makati City, Philippines.

[49] Kalyvas and Balcells (2010); Andreski (1980).

The most difficult trade-offs are required by states confronting multiple types of threats – asymmetric and symmetric, internal and external. A high-quality COIN force is needed but conventional capabilities and firepower must also be maintained. In the Philippines case, increasing encroachment by China in Philippine territorial waters underscores the need for external defense capabilities concurrent with addressing the long running insurgencies described above which adds to the complexities of the trade-offs in resource allocation and operational priorities they must make. For example, the forces capable of addressing NPA guerrilla fighters operating among the population in rural areas (asymmetric) is somewhat different than those able to storm a defended base camp of MILF separatists (symmetric semiconventional) and far different from those able to defend Philippine sovereignty against foreign aggression (symmetric conventional).[50]

Some might argue that the Philippines's decision to make investing in the types of military forces capable of addressing its internal security threats a priority, versus shifting these resources to developing capabilities to defend against external threats, result in limiting their deterrent capabilities, lowering the expected costs to the Chinese for their aggressive behavior in Philippine territorial waters and thus increasing the odds of interstate conflict in the region.

This is a trade-off the US and many of her allies must make as well. For example, fielding and maintaining a military force effective at conducting the type of protracted population-centric counterinsurgency operations pursued by US and Coalition Forces in Iraq and Afghanistan in the 2000s is not optimized for fire power intensive and maneuver-centric warfare contingencies that could arise in other theaters against other potential adversaries. Thus, states that over invest their limited resources in fielding and maintaining militaries effective at counterinsurgency risk signaling a dangerous vulnerability to potential adversaries and thus can increase their risk of symmetric interstate conflict in an anarchic international system.

War Fighting

Effectively combating insurgency can require military forces to take on a variety of missions and engage in activities outside of the bounds

[50] These distinctions in types of civil wars are introduced by Kalyvas and Balcells (2010, 201).

of those traditionally assumed by a state's military forces. Engaging in civic action programs and distributing economic aid and resources are examples, but even more extreme is the deliberate exercising of restraint in conflict to avoid civilian casualties and collateral damage in the interests of achieving operational and strategic objectives.[51] Given this, preparing military forces to conduct effective operations in a population-centric COIN campaign requires states to make trade-offs on the types of capacities developed and maintained in their military down to the small unit level. The stated mission of the US Army infantry is to "close with the enemy by means of fire and maneuver to defeat or capture him, or to repel his assault by fire, close combat, and counterattack."[52] Some argue that expanding the roles and responsibilities of the military too far outside a more narrowly defined war-fighting mandate risks effectively neutering the force and degrading its capacity to be employed in a future semiconventional or conventional threat environment. How a military fights – or does not fight – in COIN presents a trade-off at the war-fighting level for states over time. Related to the security trade-off described earlier, raising and employing forces effective against certain types of insurgent threats could potentially lead to a degradation of combat skills at the war-fighting level and reduce their effectiveness against other potential threats as a result.

Lastly, while increased civilian casualties can jeopardize political support for a counterinsurgency campaign, as described earlier, it can also impact tactical and operational level effectiveness of COIN forces at the war-fighting level. Civilian casualties can reduce information flows from the population critical for planning and initiating operational missions aimed at attrition of rebel forces as well as effective conditioning and disbursement of aid – both critical to effective counterinsurgency and how the COIN force is able to fight and operate.[53] Thus, when addressing some types of insurgent threats, states trade off operational- and even tactical-level war-fighting capabilities if they opt to employ lesser-quality COIN forces more prone to inflicting civilian casualties.

[51] McChrystal (2013); Felter and Shapiro (forthcoming).

[52] See US Army Field Manual FM 7–8 Infantry Rifle Platoon and Squad Doctrine, Chapter 1.

[53] Condra and Shapiro (2012); Berman et al. (2011); Felter and Shapiro (2017); Shapiro and Shaver (forthcoming).

Summary and Conclusions

The sources of military effectiveness in counterinsurgency cannot be adequately explained by the number of "boots on the ground" and related measures, such as the ratio of counterinsurgents to the population[54] or, in cases of third-party interventions, numbers of host nation security forces recruited and deployed to conduct counterinsurgency operations. Indeed, the qualities of the individuals filling these "boots" and how they are employed are extremely consequential in determining how these military units will perform in the field and subsequently how they will impact on the conflict.

Decentralized small units with superior leadership, training, and doctrine make the best counterinsurgents. Such units have the capacity to assess, adapt, and respond to various types of insurgency and across a wide variety of complex local threat environments. As a result, these forces perform better than other types of units operating under similar conditions against a variety of threats and by a host of measures. These findings are supported by empirical evidence from multiple distinct cases of insurgency in the Philippines as well as anecdotal evidence and experiences from other cases of insurgency.

But raising and employing the costly, high-quality military forces effective at combating insurgent threats obliges states to make a range of compromises, such as the political, security, and war-fighting tradeoffs described in this chapter, based on their prevailing and expected threat conditions. In summary, states must appreciate that the advantages in effectiveness of higher-quality military forces are pronounced in multiple ways when they are employed against asymmetric insurgent threats operating among the population but more limited when confronting semiconventional rebels or cases where government penalties for civilian casualties and collateral damage are low. Furthermore, raising and employing military forces effective in counterinsurgency can come at the expense of investing in forces capable of confronting conventional and other symmetric threats which for some states is a tradeoff that can have grave consequences if not balanced effectively based on their anticipated internal and external threat environments. Understanding the sources of military effectiveness in counterinsurgency and

[54] Quinlivan (1995–1996) describes that the ideal ratio of government forces to the population is interpreted to various extremes in military doctrine and the academic literature.

the trade-offs that must be made when investing in them is critical for all states seeking to better prepare and more effectively and efficiently employ their own military forces – or those they are supporting – to address these emerging and developed threats.

References

Andreski, Stanislav. 1980. On the Peaceful Disposition of Military Dictatorships. *Journal of Strategic Studies* 3 (December): 3–10.

Berman, Eli, Jacob N. Shapiro, and Joseph H. Felter. 2011. Can Hearts and Minds Be Bought? The Economics of Counterinsurgency in Iraq. *Journal of Political Economy* 119(4): 766–819.

Forthcoming. *Small Wars, Big Data: How the Empirical Revolution Can Help Fight and Win Today's Conflicts*. Princeton, NJ: Princeton University Press.

Biddle, Stephen. 2004. *Military Power: Explaining Victory and Defeat in Modern Battle*. Princeton, NJ: Princeton University Press.

Biddle, Stephen, and Stephen Long. 2004. Democracy and Military Effectiveness: A Deeper Look. *Journal of Conflict Resolution* 48(4): 525–546.

Biddle, Stephen, and Robert Zirkle. 1997. Technology, Civil-Military Relations and Warfare in Southern Asia. In Eric Arnett ed., *Military Capacity and the Risk of War: China, India, Pakistan, and Iran*, 317–345. New York: Oxford University Press.

Brooks, Risa A. 2003. Making Military Might: Why Do States Fail and Succeed? A Review Essay. *International Security* 28(2): 149–191.

Brooks, Risa, and Elizabeth Stanley, eds. 2007. *Creating Military Power: The Sources of Military Effectiveness*. Stanford, CA: Stanford University Press.

Callwell, C. E. 1906. *Small Wars: Their Principles and Practice*. London: Harrison and Sons.

Carey, Sabine C., Michael P. Colaresi, and Neil J. Mitchell. 2015. Governments, Informal Links to Militias, and Accountability. *Journal of Conflict Resolution* 59(5): 850–876.

Cohen, Dara Kay. 2013. Explaining Rape during Civil War: Cross-National Evidence (1980–2009). *American Political Science Review* 107(3): 461–477.

Condra, Luke N., and Jacob N. Shapiro. 2012. Who Takes the Blame? The Strategic Effects of Collateral Damage. *American Journal of Political Science* 56(1): 167–187.

Crost, Benjamin, Joseph Felter, and Patrick Johnston. 2014. Aid Under Fire: Development Projects and Civil Conflict. *American Economic Review* 104(6): 1833–1856.

Felter, Joseph H. 2005. Taking Guns to a Knife Fight: A Case for Empirical Study of Counterinsurgency. PhD diss., Stanford University.

Felter, Joseph H., and Jacob N. Shapiro. 2017. Limiting Civilian Casualties in Wartime as Part of a Winning Strategy. *Daedalus* 146(1): 44–58.

Fick, Nathaniel. 2006. *One Bullet Away: The Making of a Marine Officer*. Boston: Houghton Mifflin.

Foxworth, Lorinzo. 2012. A Grounded Theory Study of Training Transfer among Army Noncommissioned Officers. *Journal of Psychological Issues in Organizational Culture* 3(2): 30–58.

Galula, David. 1964. *Counterinsurgency Warfare: Theory and Practice*. Westport, CT: Praeger Security International.

Gates, Scott. 2002. Recruitment and Allegiance: The Microfoundations of Rebellion. *The Journal of Conflict Resolution* 46(1): 111–130.

Gentile, Gian. 2013. *Wrong Turn: America's Deadly Embrace of Counterinsurgency*. New York: New Press.

Hegre, Håvard. 2001. Toward a Democratic Civil Peace? Democracy, Political Change, and Civil War, 1816–1992. *American Political Science Review* 95(1): 33–48.

Kalyvas, Stathis N. 2006. *The Logic of Violence in Civil War*. Cambridge: Cambridge University Press.

Kalyvas, Stathis N., and Laia Balcells. 2010. International System and Technologies of Rebellion: How the End of the Cold War Shaped Internal Conflict. *American Political Science Review* 104(3): 415–429.

Kier, Elizabeth. 1995. Culture and Military Doctrine: France between the Wars. *International Security* 19(4): 65–93.

Kilcullen, David. 2006. Twenty-Eight Articles Fundamentals of Company-Level Counterinsurgency. *Marine Corps Gazette* 90(7): 50–56.

2009. *The Accidental Guerrilla: Fighting Small Wars in the Midst of a Big One*. New York, NY: Oxford University Press.

Kitson, Frank. 1974. *Low Intensity Operations: Subversion, Insurgency and Peacekeeping*. London: Faber and Faber.

Krepinevich, Andrew. 1988. *The Army and Vietnam*. Baltimore: Johns Hopkins University Press.

Leites, Nathan, and Charles Wolf Jr. 1970. Rebellion and Authority: An Analytic Essay on Insurgent Conflicts. Report RAND-R-462-ARPA, RAND Corporation, Santa Monica, CA.

Lyall, Jason. 2010. Are Coethnics More Effective Counterinsurgents? Evidence from the Second Chechen War. *American Political Science Review* 104(1): 1–20.

Lyall, Jason, and Isaiah Wilson. 2009. Rage against the Machines: Explaining Outcomes in Counterinsurgency Wars. *International Organization* 63(1): 67–106.

MacDonald, P. 2013. "Retribution Must Succeed Rebellion": The Colonial Origins of Counterinsurgency Failure. *International Organization* 67(2): 253–286.

Mao Tse-tung. 1965. A Single Spark Can Start a Prairie Fire. In *Selected Works of Mao Tse-tung*, vol. 1. Peking: FLP.

McChrystal, Stanley. 2013. *Team of Teams: New Rules of Engagement for a Complex World*. New York: Penguin.

Millett, Allan R., Williamson Murray, and Kenneth H. Watman. 1988. The Effectiveness of Military Organizations. In Allan R. Millett and Williamson Murray, eds., *Military Effectiveness*, vol. 1, 1–30. Boston: Allen and Unwin.

Mullaney, Craig M. 2010. *The Unforgiving Minute: A Soldier's Education*. New York: Penguin.

Nagl, John A. 2014. *Knife Fights: A Memoir of Modern War in Theory and Practice*. New York: Penguin.

Paget, Julian. 1967. *Counter Insurgency Campaigning*. London: Faber and Faber.

Peic, Goran. 2014. Civilian Defense Forces, State Capacity, and Government Victory in Counterinsurgency Wars. *Studies in Conflict and Terrorism* 37(2): 162–184.

Pollack, Kenneth Michael. 2002a. *Arabs at War: Military Effectiveness, 1948–1991*. Lincoln: University of Nebraska Press.

2002b. *The Threatening Storm: The Case for Invading Iraq*. Unabridged ed. New York: Random House.

Posen, Barry. 1984. *The Sources of Military Doctrine: France, Britain, and Germany between the World Wars*. Ithaca, NY: Cornell University Press.

Quinlivan, James T. 1995–1996. Force Requirements in Stability Operations. *Parameters* 25(Winter): 59–69.

Reiter, Dan. 2001. Why NATO Enlargement Does Not Spread Democracy. *International Security* 25(4): 41–67.

Reiter, Dan, and Curtis Meek. 1999. Determinants of Military Strategy, 1903–1994: A Quantitative Empirical Test. *International Studies Quarterly* 43(2): 363–387.

Reiter, Dan, and Allan C. Stam. 2002. *Democracies at War*. Princeton, NJ: Princeton University Press.

Rosen, Stephen Peter. 1995. Military Effectiveness: Why Society Matters. *International Security* 19(Spring): 5–31.

Sarkees, Meredith, and Frank Wayman. 2010. *Resort to War: 1816–2007*. Washington, DC: CQ Press.

Savage, Jesse Dillon, and Jonathan D. Caverley. 2014. American Foreign Military Training and Coup Propensity. Unpublished manuscript, MIT.

Shafer, Michael. 1988. *Deadly Paradigms: The Failure of US Counterinsurgency Policy*. Princeton, NJ: Princeton University Press.

Shapiro, Jacob, and Andrew Shaver. Forthcoming. The Effect of Civilian Casualties on Wartime Informing: Evidence from the Iraq War. *Journal of Conflict Resolution*.

Smith, M. L. R., and David Martin Jones. 2015. *The Political Impossibility of Modern Counterinsurgency*. New York: Columbia University Press.

Thayer, Thomas C. 1985. *War without Fronts: The American Experience in Vietnam*. Boulder, CO: Westview.

Thompson, Robert. 1966. *Defeating Communist Insurgency: Experiences from Malaya and Vietnam*. London: Chatto and Windus.

Trinquier, Roger. 1964. *Modern Warfare: A French View of Counterinsurgency*. London: Frederick A. Praeger.

US Army and US Marine Corps. 2007. *FM 3–24: The US Army–Marine Corps Counterinsurgency Field Manual*. Chicago: University of Chicago Press.

Valeriano, Napoleon, and Charles T. R. Bohannan. 1962. *Counter-guerrilla Operations: The Philippine Experience*. Westport, CT: Praeger Security International.

Waltz, Kenneth. 1979. *Theory of International Politics*. Reading, MA: Addison-Wesley.

6 Military Robotics, Autonomous Systems, and the Future of Military Effectiveness

MICHAEL C. HOROWITZ

When scholars and policy makers think about the utility of military robotics, they generally think about one technology, unmanned aerial vehicles (UAVs), and one use for that technology – targeted strikes against violent nonstate actors. UAVs have attained prominence due to the "drone strike" campaigns that the US has conducted over the last fourteen years in places such as Afghanistan, Pakistan, and Yemen.[1]

Viewing military robotics solely through the lens of UAVs and "drone strikes" is a mistake.[2] Current-generation UAVs are extremely limited, especially compared to their inhabited counterparts. They fly at slow speeds, making them easy to track for air defenses; they generally lack stealth capabilities, exacerbating the risk posed by air defenses; they cannot defend themselves from attack in the air; and even armed UAVs carry only limited payloads. Additionally, conducting global strikes with UAVs is a complicated organizational task requiring significant systems integration capabilities, though strikes over shorter ranges are much less complicated.[3]

Military robotics as a category, of which UAVs are a subset, is likely to spread widely to militaries around the world over the next few decades. The private-sector robotics market is rapidly expanding, with the commercial and consumer robotics market already worth tens of billions of dollars a year and growing. This underlying private-sector demand, and the fact that many advances in this arena are dual-use and

[1] Zenko and Kreps (2014a); Boyle (2015); Zenko (2014); Zegart (2016).

[2] This chapter uses the terms *uninhabited system*, *UAV*, *unmanned aerial system*, *remotely piloted aircraft*, and *drone* interchangeably, in recognition that there are significant debates over their appropriateness in different contexts. For more on definitions, see Horowitz and Fuhrmann (2017).

[3] Horowitz et al. (2016); Gilli and Gilli (2016).

based on software rather than hardware, is likely to make cutting-edge robotics technology available to many more countries than the key military technologies of the last generation, such as stealth.

This chapter examines the military robotics systems of the present and the near future to assess the implication of current trends in robotics for military effectiveness, exploring questions surrounding political and war-fighting trade-offs. Four key areas for evaluating the capabilities of military systems are their firepower, mobility, endurance, and communications. Current-generation military robotic technologies such as drones generally sacrifice effectiveness in mobility, survivability, and diversity of firepower for greater endurance and reduced risk to friendly military personnel.[4] Put another way, current-generation drones trade off overall war-fighting effectiveness for the ability to conduct long-endurance missions such as surveillance and targeted strikes. Remotely piloted military robotic systems, whether aerial drones, naval systems, or ground robots, also generate a secondary set of war-fighting trade-offs with their vulnerability to hacking, spoofing, and other cyber-based threats, which could limit their effectiveness and usage.

In particular, the vulnerability of remotely piloted military robotic systems to hacking, spoofing, or the denial of access to communications could place an inherent limit on their utility. Addressing the challenges of hacking and bandwidth could also create incentives to shift from remotely piloted systems to systems with greater degrees of machine autonomy – relinquishing some human control over the use of force for the promise of more effectiveness. Lethal autonomous weapon systems (LAWS) can select and engage targets on their own, and could appear to help militaries hedge against the risk of hacking by advanced adversaries with significant cyber and electronic warfare capabilities. Yet, delegating more control to machines is a trade-off itself, and one with important moral and ethical, as well as military, consequences.

The chapter proceeds by laying out what military robotics are and their anticipated trajectory. While early forms of military robotics have been around for decades, their rapid growth in the last decade makes laying out their likely development pathways a relevant task. A discussion of the technology is necessary because it not only helps frame the discussion of trade-offs that follows, but illustrates uncertainties about the development of the technology.

[4] The impact of drones on civilian casualties is discussed below.

The chapter then introduces a framework for thinking about the effectiveness of new military technologies, before evaluating the specific effectiveness of current-generation military drones as well as other current-generation military robotics. The chapter concludes by describing how some of the key trade-offs involved in the use of remotely piloted military robotics could generate incentives for the introduction of more machine autonomy into weapon systems.

What Are Military Robotics?

Military robotics represent the fusion of advances in robotics with military hardware. Most military robotic systems (MRS) are used in missions where they substitute for inhabited platforms. For example, current-generation UAVs are generally used as substitutes for manned aircraft, whether the mission is surveillance or strike. Military robotic systems are remotely piloted – or at least remotely supervised in-flight – with decisions about the selection and engagement of targets made by human soldiers in response to the sensor inputs they receive from the platform.[5] Thus, the use of force is still under direct human control with armed military robotic systems today, albeit potentially over great distances (though that is incidental to the definition). It is also possible, however, to develop military robotics that could operate autonomously, beyond human control after activation.[6] Figure 6.1 highlights one way to think about the relationship between robotics and autonomy.

Robotics technology is spreading around the world for both military and commercial purposes. In the military arena, a conservative estimate by Global Industry Analysts Inc. suggests that defense spending on robotics will reach $7.5 billion a year by 2018.[7] While far less than the amount spent on manned systems, it represents a dramatic increase. Other projections suggest that even larger growth is on the way. A related study by Global Industry Analysts Inc. projects

[5] The RQ-4 Global Hawk has some autonomous flight features once a pilot on the ground programs in a destination, but that is more akin to autopilot.

[6] Given that LAWS generally do not exist today (outside of limited exceptions such as the Israeli Harpy), and significant controversies remain on the definitional front, a full assessment of LAWS is beyond the scope of this chapter. Also beyond the scope of this chapter is the development and use of military robotics by militant groups such as Hezbollah, though that is an important topic for future research (Bergen 2014).

[7] Global Industry Analysts (2012).

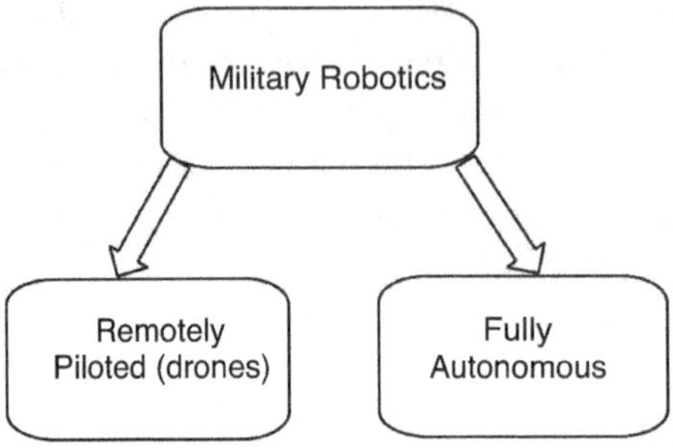

Figure 6.1 Relationship between military robotics and autonomy.

the mobile robotics market, both civilian and military, will exceed $14 billion a year by 2019.[8] Estimating the size of the military and commercial UAV market, a subset of robotics, the Teal Group predicted in 2014 that spending will almost double from $6.4 billion in 2014 to $11.5 billion a year by 2023.[9] In agriculture, the need for robotics to monitor, water, and potentially even harvest crops could lead to an annual market worth about $16.3 billion a year by 2020, according to Wintergreen Research.[10]

These estimates of the robotics industry may underestimate the way the robotics market will grow over the next decade. Existing market estimates mostly focus on industrial robotics and other relatively stable growth areas. With developments in mobile technology, and especially service robotics, combined with continued growth in programming for some types of artificial intelligence, growth in the robotics market over the next decade could exceed existing industry estimates.[11]

Advanced and armed UAVs have spread around the world just in the last five years. About seventeen countries now have armed UAVs, up from seven just eighteen months ago. This is not just a trend limited to the most advanced economies in the world. Countries such as Nigeria, Pakistan, and Iran reportedly now have armed drones potentially capable of conducting at least short-range operations.[12] Countries are interested in military robotics for a variety of missions. UAVs could assist with delivering emergency assistance in disaster relief operations. UAVs could also help locate and track migrants in dangerous situations – and

[8] Transparency Market Research (2014). [9] Teal Group Corporation (2014).
[10] Carroll (2014). [11] Bloomberg Business (2012).
[12] Fuhrmann and Horowitz (2017).

are already being employed to help track migrants fleeing Syria and crossing the Mediterranean to Europe.[13]

China is investing heavily in military robotics as part of its attempt to build military capabilities that will help it compete with the US. China's heavy investments in UAVs and military robotics have attracted significant attention. The Defense Science Board in the US in 2012 stated, "The military significance of China's move into unmanned systems is alarming."[14] China's debut of the Wing Loong at the 2013 Paris Air Show signaled its competitive entry into the burgeoning UAV export market, while its November 21, 2013, test flight of the Lijian, an alleged stealth UAV, demonstrates that Chinese developments of military robotics are not simply at the "low end" of technology.[15]

After a decade of mostly sitting on the sidelines of developments in military robotics, Russian investments have also grown. Moscow has discussed deploying armed ground robots to conduct perimeter patrols at its ballistic missile bases and long-endurance UAVs capable of monitoring potential competitors in the Arctic.[16] Yuri Borisov, the Deputy Minister of Defence in Russia, stated in 2014 that R&D on an indigenous Russian armed UAV is close to complete, with flight tests scheduled to start in 2017.[17]

Especially for countries with smaller populations and advanced economics, robotics could be simply the next step in the increasing focus on capital, as opposed to labor, in building modern militaries. European countries, increasingly dependent on the US for power projection and facing high labor costs, have also shown great interest in military robotics. In May 2015, for example, Italy, France, and Germany announced a joint plan to develop a next-generation military UAV.[18] Italy and France have already imported unarmed MQ-9 Reapers from the US.

Developments in the Asia-Pacific further highlight growing interest in military robotics. These investments demonstrate how the dual-use nature of robotic systems, and the fact that the critical underlying components are much more hardware than software, mean military robotics are likely to spread much faster than more purely military

[13] Miller (2015). [14] Defense Science Board (2012, 71).
[15] Office of the Secretary of Defense (2014, 33); BBC News (2013); Bender (2015).
[16] Jones (2014); Wong (2014). [17] Litovkin (2014).
[18] Kington and Tran (2015).

technologies such as stealth technology, which does not have much if any commercial application.[19] Japan recently purchased the RQ-4 Global Hawk from the US,[20] while South Korea developed and deployed the SGR-A1, an armed system with at least some autonomous capabilities, along the DMZ.[21] In Southeast Asia, Dr. Ng Eng Hen, Singapore's Minister of Defence, stated in 2014 that Singapore was thinking hard about the capabilities for its future soldiers, including micro-UAVs for individual soldiers and robotic mules to carry equipment.[22] Taiwan is also considering UAVs for its air force.[23]

Essentially, many countries are already converting their expertise in private-sector robotics into new military investments. Military robotics are therefore critically important to understand if for no other reason than lots of countries are rapidly investing in MRS across a variety of military domains.

Historical Analogies for Military Robotics

Military robotic systems have some conceptual similarities to emerging new military technologies of the past. All new military technologies face initial periods of uncertainty regarding their effectiveness and raise organizational issues surrounding their employment.[24]

Some new technologies end up integrated into militaries around the world, such as the tank or, more generally, the combustion engine, while some do not, such as the neutron bomb. Integration requires at least some mechanical and organizational learning.

Ballistic missiles and cruise missiles have some conceptual similarity to military robotic systems. Missiles were initially seen as a potential substitute for fighter/bombers because they could deliver payloads over large distances, theoretically generating a similar level of military effectiveness to airpower, but without putting their friendly troops at risk from air defenses. They also have some regulatory similarity, since the Missile Technology Control Regime (MTCR), the international agreement covering the export of missile technology, incorporates drones as well.

About thirty countries have ballistic missiles of some type, with "ranges from about 80 km to more than 10,000 km."[25] During the

[19] Horowitz (2010). [20] Hardy (2014). [21] Prigg (2014). [22] Hen (2014).
[23] Axe (2011). [24] Horowitz (2010). [25] Mettler and Reiter (2013, 856).

Cold War, countries developed ballistic missiles to avoid putting their own soldiers at risk for those strike operations, and ballistic missiles are hard to shoot down, especially given the vulnerability of missile defense systems.[26] Finally, ballistic missiles can reach targets faster than many other existing kinetic weapon delivery systems, making them especially useful for rapid attack missions.[27]

The proliferation pattern for ballistic missiles potentially offers some lessons for military robotic systems such as UAVs. Higher-end ballistic missiles have proven much harder to design than anticipated. The guidance challenges at beyond visual range remain complex, as is miniaturizing explosives for all but the most advanced defense sectors.[28]

Drones have already spread far wider than ballistic missiles. About ninety states had acquired military drones by 2015. While the basics of ballistic missile technology are dual-use, meaning even less advanced economies have the potential to develop and use basic ballistic missiles, that is even more the case with drones. As many countries (about thirty) have advanced drones today (excluding the most advanced global strike UAV platforms) – as have acquired ballistic missiles over several decades.

Military robotics may share more in common with cruise missiles than ballistic missiles. Cruise missiles follow flatter, nonballistic trajectories and generally have midcourse guidance options, such as GPS. The potential for midcourse guidance, which military robotics systems have as well, conceptually, is one reason why drones were initially grouped under the MTCR.[29] At that point, drones meant target drones used to test the accuracy of other weapon systems, not weapon platforms in and of themselves.[30]

In 1959, the Soviet Union deployed the first mass-produced cruise missile, the SS-N-S Styx, which was then manufactured by China and North Korea as well.[31] Seventy countries possessed some type of ground or sea launched cruise missiles by 2000.[32] About twenty countries possessed air-launched cruise missiles – generally considered more

[26] Freedberg (2015).
[27] This list draws from Mettler and Reiter (2013, 856–859).
[28] Karp (1996). [29] Missile Technology Control Regime (1987).
[30] Loitering cruise missiles such as the Harpy are not exactly drones but not exactly traditional cruise missiles. The Harpy is discussed in greater detail below.
[31] Carus (1992, 13); Mahnken (2005, 9).
[32] General Accounting Office (2000, 5).

technically challenging – in 2005,[33] and those numbers have held relatively constant over the last decade, though the capabilities of those countries, such as China, have continued growing.[34]

Point-defense systems such as Israel's Iron Dome system, the US-deployed C-RAM, the US-developed SM-6, and other platforms offer potential defenses against the cruise missiles of today, as opposed to ballistic missiles, though the potential for hypersonic cruise missiles could undermine the effectiveness of today's cruise missile defenses.[35]

Neither ballistic missiles or cruise missiles have replaced strike aircraft as a tactical weapon in conventional conflicts (though ballistic missiles have mostly displaced bombers for carrying strategic nuclear weapons, however). First, excluding basic short-range cruise missiles that simple defense systems can often shoot down, the engineering challenges necessary to manufacture advanced versions were higher than initially anticipated. Constraints on the export market after the MTCR came into effect and states such as China began following the MTCR in the late 1990s also slowed acquisition.

Relatively high unit costs also limit the use of advanced missiles. For example, even the US military has a limited inventory of Tomahawk missiles because its control and reliability comes with a large financial price tag. One estimate suggests that current-generation Tomahawk missiles cost $569,000 each in 1999 US dollars, while GPS-guided Joint Direct Attack Munitions launched from attack aircraft, even with advanced laser guidance options, cost only $42,000 each.[36]

Finally, attack aircraft maintain some accuracy and controllability advantages – especially in comparison to ballistic missiles.

The history of ballistic and cruise missile proliferation would suggest a future for military robotics where they serve as a supplement for many militaries, but higher-end capabilities are limited to the most powerful militaries.

An alternative historical analogy might suggest another path, however. There are some potential similarities between military robotics and the introductions of new technologies such as submarines in the late nineteenth century and the tank, airpower in general, and the

[33] Mahnken (2005, 13). [34] Farley (2014).

[35] *Economist* (2013); Cohen (2015).

[36] US Navy (2014); Kasper (2015). In comparison, a Hellfire missile launched by an MQ-9 Reaper costs about $100,000 per missile (Kasper 2015).

aircraft carrier in World War I. In each case, there was an initial demonstration of the technology, followed by extensive discussions of the technology's limitations in comparison to current systems.[37] If countries had ceased investing based on their limitations compared to previous ways of generating force, they would have missed several critical breakthroughs in the use of military force as the technologies matured and were joined with novel operational concepts.

Evaluating the military effectiveness of airpower, for example, based purely on its performance in World War I and early testing in the 1920s, would have significantly underestimated its promise. After all, one of the leading British fighters in World War I was the Airco DH.2, a biplane with a ceiling of 14,000 feet, range of 250 miles, and one machine gun. By World War II, the British had multifaceted aircraft, including bombers, dive bombers, fighters, and surveillance aircraft. The British Supermarine Spitfire had a ceiling of 36,500 feet, a range of 470 miles, and was armed with several machine guns.

Similarly, today, there are advocates of military robotics who argue military robotic systems could play a substantial role in the militaries of the future, and critics who point to the limitations of the technology. Appropriately assessing the military effectiveness of MRS therefore requires not just analyzing today's capabilities, but those that appear likely in the future as well.

The Effectiveness of Military Robotics

As described above, military robotic systems are spreading across national militaries in the ground, naval, and air domains. Moreover, many of these systems are experimental or early in their life cycle. Thus, any assessment of their effectiveness – and the war-fighting trade-offs they both solve and induce, is necessarily tentative.

Consider three possible dimensions of military robotic systems: operational status (whether they exist already or are in development), operating area (air, naval, or ground),[38] and operational concept (will they be used as 1:1 replacements or supplements for similar manned systems, or do they require more significant organizational changes to

[37] On airpower, see Horowitz and Reiter (2001).

[38] While there could be military robotic systems in space, there are none at present, so it is not represented in the table.

Table 6.1 *Military Robotic Systems Matrix*

	Current generation	Next generation	
	Existing Operational Concepts and Systems	*In-Development Operational Concepts and Systems*	*Potential Future Operational Concepts and Systems*
Air	Surveillance and strike in permissive environments (MQ-9 Reaper – US)	Surveillance and strike in nonpermissive environments (Sharp Sword – China)	High-altitude Surveillance (SolarEagle – US)
Ground	Border patrol (Guardium – Israel)	Surveillance and patrol (Wolf-2 – Russia)	
Naval	Coastal patrolling (Protector – Israel)	Mine hunting/disposal (Knifefish – US)	Suicide drone boats (Iran)

use). Table 6.1 describes this matrix, including the names of some current and planned platforms to illustrate the categories.

The most prominent area of military robotics at present involves the use of aerial drones, meaning the operational status is the present, the operating area is the air, and the robotics are typically used as 1:1 replacements or substitutes. Thus, this section assesses the military effectiveness of this most common current generation of aerial drones, along with next-generation aerial drones already on the drawing board and under development, discussing key war-fighting tradeoffs that drive their design and use. It then turns to looking at the effectiveness of military robotic developments in the naval arena. Finally, this section explores the effectiveness implications of different potential operational concepts for using military robotics, such as swarming.

Drone Military Capabilities, Present and Future

International relations research evaluating the effectiveness of military technology, such as research on the offense–defense balance, tends to focus on the relative capacity of military technologies to aid

Table 6.2 ***Elements of Military Platform Effectiveness***

Platform effectiveness element	Core concept
Firepower/lethality	The ability to accurately target with relevant munitions
Mobility/vulnerability	The speed, maneuverability, and stealth capacity of a platform
Endurance	The duration a platform can sustain itself for
Communications	The sensor and transmission capabilities of the platform

overall operations by providing mobility or generating firepower.[39] These broad categories contain subelements, including platform speed, stealth, and self-defense capacity (which influences survivability), platform endurance, and platform accuracy. These facets are related; mobility is a function of both speed and stealth, which influences vulnerability, but they have enough distinct components that they are worth evaluating independently. Table 6.2 lays out the four facets of effectiveness evaluated below.

How do current and in-development UAVs compare to the most likely substitutes for those capabilities in relevant missions, e.g., for a drone surveillance or drone strike mission, how would the relevant drone compare to inhabited attack aircraft?

Firepower

The most advanced deployed armed drone in the world, the MQ-9 Reaper employed by the US, can carry four air-to-ground (AGM)-114 Hellfire missiles, two GBU-12 Paveway II laser-guided bombs, or two GBU-38 JDAMs.[40] Compared to attack aircraft such as the F-18 or F-15, this is an extremely limited payload in scope and volume.[41] This limited weapons payload also gives it much less utility than the A-10 in a close air support role, or even many attack aircraft not designed for close air support. Thus, in the firepower dimension, using a drone

[39] Lieber (2000); Jervis (1978); Van Evera (1999).

[40] US Air Force (2015). There have been operational challenges in actually employing JDAMs from MQ-9 Reapers, however, making this capability more notional.

[41] US Navy (2009).

involves a trade-off between reducing the risk to your military personnel – who no longer have to fly above adversary air space – and the firepower of the platform.

UAVs do have potential advantages in their ability to strike precisely against small groups or individuals, however, due to their ability to loiter over a target. The higher endurance of UAVs, as described below, provides more situational awareness to pilots, giving them a greater ability to verify targets than an attack aircraft pilot with less airtime over the target.[42] The real-time situational awareness provided by loitering also allows drone pilots to be more responsive and divert missiles up to the last moment if something changes on the ground.[43] This greater awareness means the use of UAVs likely reduces noncombatant deaths in addition to increasing the likelihood of hitting the target when the strike option is a UAV versus another type of military technology.[44]

Next-generation UAVs will likely narrow the gap between manned systems and drones. China is developing the Lijian, or Sharp Sword, while the US Navy developed a test-bed platform called the X-47B, though it did not lead to an armed procurement program at present.[45] Some sources suggest, however, that if the US Navy built the most extensive next-generation platform possible, based on lessons from X-47B testing, it would hold 85 percent of the weapons load of the new F-35 Joint Strike Fighter.[46] Thus, there is some evidence that, in theory, drones in development could rapidly reduce the gap between the weapons capabilities of inhabited aircraft and drones.[47]

Mobility

Current-generation drones lag significantly in speed and stealth compared to manned attack aircraft. The MQ-9 Reaper only cruises at about 200 knots, compared to an aircraft such as the F-18, which can reach speeds of up to Mach 1.7+, or over 1,124 knots.[48] Moreover, the shape of current-generation drones such as the MQ-9, with many sharp angles, makes them easy to observe with radar. This restricts their use

[42] Davis et al. (2014, 11–12); Horowitz et al. (2016).
[43] Zenko and Kreps (2014, 8–9).
[44] Lewis (2013). The question of drone strikes and civilian casualties is covered in more detail below.
[45] Defense Update (2013). [46] Eaglen and McGrath (2015).
[47] Political will to produce drones with those capabilities is a different story, of course.
[48] US Navy (2009).

to permissive environments – locations without air defenses or with air defenses shut down.[49] Even countries with rudimentary air defenses or older fighters have the capacity to shoot down today's advanced drones. For example, in 2013, when the government of Iran reportedly scrambled their F-4 Phantom fighters after they detected an MQ-1 Predator nearing Iranian air space, the US launched its own fighters to protect the Predator and ensure it got home safely.[50]

Some surveillance drones, such as the RQ-170 Sentinel in the US, reportedly have some stealthy capabilities. The batlike wingspan of the X-47B and Sharp Sword, however, suggest that higher-end next-generation drones could become more stealthy. Technically, drones could also offer even greater speed and maneuverability potential than manned aircraft. Manned aircraft have limits on maneuverability due to the G forces the human body can absorb.

Current-generation drones also lack air-to-air capabilities or countermeasures to respond to airborne threats, something that both manned fixed-wing aircraft and helicopters possess. This further limits their potential usage to permissive air space, but there is nothing technical that would prevent next-generation systems from including self-defense capacity.

Endurance

While mobility is a significant disadvantage for current-generation drones, endurance is a significant advantage. An F-18 E/F conducting a combat air patrol from an aircraft carrier has an effective mission time of about two hours and fifteen minutes according to some estimates, far less than the thirteen hours of a fully armed MQ-9 Reaper.[51] While an F-18, along with other advanced attack aircraft, can be refueled in the air, giving them more potential time on station, they still face fundamental time limitations due to the endurance of the human in the cockpit.

Expanding to comparisons of surveillance aircraft, the RQ-4 Global Hawk Block 40, a US surveillance drone, flies at sixty thousand to sixty-five thousand feet and can loiter around a target at only 310 knots TAS (True Air Speed) for up to twenty-four hours.[52] The closest comparison is the U-2, a manned surveillance airplane flown by the

[49] Davis et al. (2014). [50] Cenciotti (2013).

[51] Federation of American Scientists (2000); Wheeler (2012). Without weapons, an MQ-9 Reaper can stay in the air for almost thirty hours.

[52] Northrop Grumman (2012).

US military since 1957. It flies at seventy thousand feet, but even with refueling, its maximum endurance is about twelve hours.[53]

Next-generation drones may lose some endurance as they gain weapons and other capabilities, which would make them heavier. Future UAVs could have air-to-air refueling capabilities as well.[54]

Adding air-to-air refueling capabilities to drones makes their potential duration more about mission requirements than technical capacity, since they could stay in the air for days without the limitation of having to land to switch out pilots.

Communications

Communications here refers to sensor packages that give the pilot – whether on the plane or thousands of miles away – the information necessary to conduct strikes or conduct surveillance. There is rough equivalence between drones and manned attack aircraft sensors – for those missions drones can conduct. Of course, for missions such as air-to-air combat or other long-range strikes, manned aircraft carry sensor pods and radar capabilities that drones do not have. However, for air-to-ground missions, the sensor pods on an MQ-9 Reaper, which give the pilot real-time satellite connectivity and imagery data, are essentially equivalent to those on armed attack aircraft conducting similar missions.

In the surveillance realm, due to its larger size and heavier weight, the U-2 can carry a larger array of sensor pods than the RQ-4 Global Hawk for any given mission. This gives the U-2 a potential advantage over the RQ-4 Global Hawk in terms of its diversity. Continued shifts within advanced militaries toward more uniform intramilitary sensor pods means any gap between inhabited and uninhabited aircraft will likely decline over the next several years.[55]

Implications for Drone Effectiveness

Overall, current-generation drones thus trade lower levels of firepower, mobility, and survivability for greater endurance, control over munitions until the moment of impact, and reducing the risk of

[53] Northrop Grumman (2012); Lockheed Martin (2015); Michel (n.d.).
[54] LaGrone (2015).
[55] Not that each deploys the same sensor platforms but that each will deploy a smaller number spread across a larger number of its own platforms.

Table 6.3 ***War-Fighting Trade-offs When Substituting Current-Generation UAVs for Manned Attack Aircraft***

Benefit of using UAVs	Cost of using UAVs
Reduce risk to friendly forces	Less mobility
More precise control over the platform and munition	Less survivability
Greater endurance	Smaller and less diverse weapons payload

friendly casualties (except via friendly fire) to zero. There are war-fighting (endurance) and political (friendly casualties) rationales for the trade-off.

Especially given the casualty sensitivity of the public in democracies, even though the risk to pilots in permissive air space such as Afghanistan and Pakistan is extremely low, reducing that risk to zero through the use of drones creates political benefits for democratic leaders.[56]

Survey evidence in the US suggests broad public support for drone strikes. Survey experiments further demonstrate that the lower risk to US military personnel is a large part of what drives support for drones in the US, illustrating this political trade-off in action.[57] For democratic polities where capital is more abundant than labor, leaders may be willing to accept more constrained missions to avoid putting their soldiers at risk. This could be one reason why, in 2015, European countries such as Spain and the Netherlands pursued advanced drones.[58] Table 6.3 lays out the overall trade-offs discussion in this section more clearly.

The promise of greater control over the use of military force may also make countries willing to accept the reduction in overall military effectiveness relative to manned attack aircraft (especially if they believe that the technology is rapidly improving). For democracies and mixed regimes, high-level military and political leaders often seek direct control of the battlefield – such as Nixon picking targets in the Vietnam

[56] Davis et al. (2014); Zegart (2016).

[57] Horowitz (2016); Kreps (2014); Cillizza (2013). Kreps shows that when the public is given additional information on the legality and effectiveness of drone strikes, support does decline.

[58] Villarejo (2015); Defense Security Cooperation Agency (2015).

War – even though decentralized operations are generally more successful at the small unit level.

Control over far-flung military forces also provides a strong motivation for autocratic development and acquisition of UAVs (and military robotics in general). Autocracies are more likely than democracies to face civil wars, insurgencies, and lower-level strife. Autocratic leaders therefore rely on remotely deployed troops to conduct repression, with some risk that distant troops will not carry out orders. Many autocracies fall, after all, precisely because the military decides it will no longer engage in repression of the population.[59]

Armed UAVs could make therefore "remote control repression"[60] possible for some autocratic leaders. Instead of needing to trust distant military commanders, autocratic leaders can centralize more military power in the hands of a smaller portion of their selectorate. This domestic use of drones for surveillance and strike purposes is not just theoretical. China used its drones in August 2014 in Xinjiang after riots and protests erupted.[61] With territory ranging from Xinjiang and Tibet to the South China Sea, UAVs appear useful for the central government in Beijing to monitor its periphery.

Former US Deputy Secretary of Defense Bob Work recently argued that autocracies, because they do not trust their own populations, are especially likely to aggressively invest in military robotics and automation to increase their direct control over the use of force.[62] UAVs (as well as military robotics as a category) could therefore potentially help autocracies avoid coup-proofing strategies such as weakening their military for external conflicts to maximize its effectiveness domestically.[63] Autocrats worried about elite threats can keep a closer eye on those elites, and restrict their independent operating authority. Autocrats can station UAV pilots at central locations guarded by their most trusted troops. Of course, there is a potential trade-off here as well – centralizing force means that if the autocrat ever loses control over UAV operations, it could place their rule at risk much more quickly.

Drones and Financial Costs

Another potential trade-off (political as well as war fighting) that military robotics are already generating has to do with financial costs. The

[59] Davenport (2007); Svolik (2013). [60] Cronin (2013). [61] Tatlow (2014).
[62] Work (2015). [63] On coup-proofing, see Talmadge (2015).

MQ-9 Reaper costs about $16 million in 2006 US dollars, or $64 million for four, the number used for a twenty-four-hour combat air patrol. In contrast, the average projected unit cost for an F-35 (averaged across the A, B, and C models) is $135 million, and the program is behind and overbudget, meaning costs could increase. While the F-35 is dramatically more capable in nearly all ways than the MQ-9, it only has a fraction of the endurance.[64]

When the logistical tail of current US UAV operations are taken into account, however, the cost savings compared to manned systems are not as large as one might think, but they do likely exist.[65] Data are still limited on tooth-to-tail cost comparisons, but one of the only true apples-to-apples comparisons of unmanned systems with manned counterparts, the US Air Force comparison of the U-2 spy plane and the RQ-4 Global Hawk, concluded that the RQ-4 could accomplish the same missions at a lower price point.[66]

Nevertheless, this trend is likely to make the proliferation of military robotics more likely. If it is possible for militaries to acquire cutting-edge technology, even at smaller purchase levels, they are often likely to do so for reasons that have to do with prestige or interoperability as much as building their direct military capabilities.[67] In the period after World War II, for example, one of the reasons why so many countries such as Canada and the Netherlands decided to completely bandwagon with the US Navy and decommission their aircraft carriers had to do with the excessive unit costs required to operate those carriers.[68] If the relative unit costs had been lower, perhaps the situation would have been different.

Dual-use software could bring down unit costs further. Minimally manned or completely unmanned naval ships might have lower unit costs than manned counterparts, for example, because they do not have to be designed to build in survivability for crew, potentially freeing them up to store more munitions, build the capacity for greater speed, or other potential improvements.

Drones, Civilian Casualties, and Blowback

Finally, there is the question of whether drones make collateral damage, in the form of civilian casualties from drone strikes, more likely, which

[64] Cohen (2015); Gertler (2014). [65] Wheeler (2012).
[66] Everstine (2014); Robson (2014). [67] Goldman and Eliason (2003).
[68] Horowitz (2010).

could introduce independent security trade-offs and undermine the logic of using drones for counter-terrorism missions. As described above, in comparison with strikes from other sources, the loitering and situational awareness of drones make them relatively less likely to generate noncombatant casualties.

The use of drones for targeted strike missions, however, means drone strikes are often associated with missions that lead to civilian casualties, leading to a perception that drone strikes generate more civilian casualties than other weapons.[69] Empirically testing for a relationship between drone strikes and civilian casualties requires additional data that are not available at present. First, researchers need data on non-drone strikes in the same conflicts in which drone strikes occur, including civilian casualty data. Second, for each strike, drone or no drone, researchers need to know the counterfactual options. In some cases, a drone strike may occur because the other platforms in the relevant geographic area are not carrying the appropriate munition for a strike.

For example, the AGM-114 Hellfire missile fired by the MQ-9 Reaper is also employed by the Apache Helicopter. If a strike is necessary in a location where there is not an Apache, it is not drone strikes per se that are the issue as much as military necessity and the only available platform being a drone. Essentially, it is necessary to separate the platform from the munition, since the particular munitions being carried at a given point, along with timing, can influence the selection of the platform for the mission.

In other cases, if the counterfactual option for a drone strike was no strike at all, then any civilian casualties generated through that strike would be additive, e.g., civilian casualties due to drone strikes. Without additional data, drawing inferences about the relationship between drone strikes and civilian casualties is extremely difficult. With the release and declassification of additional data over time, hopefully researchers can test these claims.

There is also the risk that drone strikes lead to blowback, inspiring people in targeted areas to become insurgents.[70] This would be a classic security threat trade-off where the use of military force in one arena can lead to negative externalities in another arena. Recent research by Johnston and Sarbahi, however, suggests that this risk may be overstated. Using data on drone strikes in Pakistan from 2007 to 2011, they

[69] Ackerman (2014). [70] Cronin (2013).

show that drone strikes actually make terrorist attacks in targeted regions less likely.[71] These results are only from one location and not necessarily long term. In addition, the results do not preclude the possibility that drone strike operations in one part of the world could boost recruitment for similar insurgent groups in another part of the world or inspire lone wolves such as the failed Times Square bomber, Faisal Shahzad.[72]

Operational Planning for Military Robotics: Impact on Effectiveness

When it comes to potential military effectiveness of new technology categories, it is always tempting to think about them as direct replacements for what preceded them. The tank replaced mounted cavalry, while the aircraft carrier replaced the battleship. The reality is that while those replacements occur at the surface, military technology transitions often involve significant shifts in operational concepts as well. As motorized vehicles, tanks could move faster and at greater distances than mounted cavalry, giving them greater striking power en masse, though they sacrificed something in mobility since they were more limited in the type of terrain they could traverse (especially early-generation tanks). Tanks had their biggest impact when they enabled operational strategies such as blitzkrieg that were simply not possible with cavalry. Aircraft carriers were never a 1:1 replacement for battleships – a single aircraft carrier could generate significant striking power since the carrier was simply a mobile air field, meaning it operated completely differently from the battleships of World War I.

While some ways of integrating new capabilities require higher levels of organizational transformation to implement, making them less likely for many actors, other types of transformation can occur with much less difficult organizational change.[73] Military robotics are not an exception. There are several way to think about the integration of military robotics into modern militaries – 1:1 replacement, many-to-one replacement, man–machine teaming, and swarming – and each has different potential organizational consequences and thus implications for military effectiveness.

[71] Johnston and Sarbahi (forthcoming). [72] Adams and Nasir (2010).
[73] Horowitz (2010).

The least disruptive pathway for the further integration of military robotics into modern militaries, aside from military robots with the capacity to conduct new missions, is 1:1 replacement for manned systems. These are situations where a country has some number of trucks, fighters, cargo jets, or something else and replaces them with uninhabited systems that accomplish the same task in the same way. This is organizationally disruptive for those pilots or drivers displaced, but it is less organizationally disruptive in terms of doctrine and operational concepts, since the general methods for using force remain more or less the same. The 2013–2014 budget debate in the US Air Force about whether to keep the U-2 or Global Hawk, described above, conceptually demonstrates debates about 1:1 replacement in action.

A second way to think about the integration of military robotics involves many-to-one replacement. In this scenario, unmanned systems are designed to conduct the same missions, but with many lower-quality systems replacing fewer higher-quality manned systems. These would not be systems necessarily designed to "swarm." Rather, this would represent a shift from a focus on quality in military platforms, which has dominated military procurement in the US and other major powers since the 1970s, to quantity. The diffusion trends discussed above could accelerate this trend. If dual-use software and commercial robotics make it easier for more countries to build systems of similar quality, quantity could become a more important means by which militaries attempt to differentiate themselves.[74] This would be more bureaucratically challenging for many militaries than 1:1 replacement because it involves different production models and concepts of operation. It would be more of a return to mass-based operational concepts from the nineteenth century than the quality-based operational concepts of the last few decades.

A third mechanism for the further integration of military robotics involves manned–unmanned teaming. In this model, manned assets would lead and coordinate unmanned assets in-theater. For example, imagine a manned F-35 surrounded by multiple, older-generation F-18s (or other fighters or drones) piloted either by the F-35 pilot or remotely controlled from distance. This could magnify the overall striking power of a sortie, allow the manned platform to direct an unmanned platform to take greater risk when facing potential air

[74] Horowitz (2014); Work and Brimley (2014).

defenses, and could lead to interesting tactical innovations. Existing experimentation suggests some militaries are already thinking along these lines.[75] For European countries with smaller militaries, this could be a promising potential way to leverage their human capital.

Teaming could be a bureaucratically more palatable way forward, as well, since it keeps pilots and drivers at the center of the activities of the military services. It does not, though, eliminate the risk of hacking or spoofing.[76] In some ways, it exposes military personnel more directly to those risks, since they would presumably be in reasonably close proximity to robotic systems. Local teaming could, however, alleviate some of the global bandwidth constraints and dependency on satellite networks, making the effective use of military robotics easier.

The final, and least understood, mechanism for thinking about the integration of military robotics into modern militaries involves developing numerous systems that work together in semiautonomous or even autonomous fashions, or swarms. In this model, while the emphasis is quantity, rather than quality, the operational concepts involve devolving more control to computerized coordination between massed platforms rather than individual remote piloting. For example, in early 2015, Lt. Gen. James Holmes, US Air Force (USAF) Deputy Chief of Staff for Strategic Plans and Requirements, referenced the notion of autonomous swarms as a potential conceptual alternative for the USAF as it imagines the design parameters for a next-generation fighter.[77]

Naval Robotics

As highlighted at the outset of the chapter, there is also growing investment in military robotics outside of the UAV realm. One example is investments by navies. The US Navy is working on unmanned boats with at least some autonomous guidance that would have the ability to "swarm" potential adversaries.[78] Part of the logic behind these systems involves lower unit costs. According to the former Chief of Naval Operations in the US Jonathan Greenert, "this networking

[75] Osborn (2015).

[76] Spoofing is when a system is fooled into thinking something is happening that is not really happening, like convincing a plane to put down its landing gears because it thinks it is near the ground even though it is thirty thousand feet in the air.

[77] Copp (2015). [78] Smalley (2014).

unmanned platforms demonstration was a cost-effective way to integrate many small, cheap, and autonomous capabilities that can significantly improve our warfighting advantage."[79] Interestingly, Iran has been working on similar programs, with Iran's reputed Arians suicide drone boat a potential tactic to counter US naval superiority in the Strait of Hormuz.[80] Israel has already deployed an armed unmanned surface naval vessel called The Protector to remotely patrol Israel's coastline. Since it is unmanned, the ship has extremely high maneuverability, not having to be concerned with personnel going overboard or otherwise getting injured.[81] It demonstrates one way that coastal navies may be able to bring down overall costs, since in regular operations, fewer personnel may be necessary for operations than manned patrol ships.

Unmanned underwater vehicles (UUVs) could also have a variety of uses. The US Navy used the Bluefin-21 as part of the search for Malaysia Airlines Flight 370. Robotic systems could also substitute for dangerous naval missions such as sweeping for sea mines. The US Navy's Knifefish program is designed to autonomously sweep for potential mines in littoral combat areas.[82] Given the opacity of the ocean, these systems, though not armed, are programmed to act relatively autonomously, and could offer navies significant value at low cost – and low risk to personnel. A RAND study of Chinese military robotics research in 2015 suggests that China's interest in UUVs has also "increased dramatically,"[83] likely for surveillance purposes given China's large number of territorial disputes.

Broader War-Fighting Trade-offs with Military Robotics: Hacking and Bandwidth

In addition to the issues described above, there are additional military effectiveness issues with military robotics that are potentially inherent to the category. Most broadly, information security concerns could limit the utility of military robotics. Military robotics require not just large amounts of bandwidth, but secure access to that bandwidth. Just as hacking and spoofing are huge challenges in the commercial realm, they are in the military arena as well. Remotely piloted military

[79] Smalley (2014). [80] Peck (2015). [81] Orpaz (2013). [82] Tadjdeh (2014).
[83] Chase et al. (2015, 3).

systems operate, conceptually, in ways similar to remote "smart home" technologies such as Nest. This makes them vulnerable to hacking – and there is already a thriving cottage industry of hackers who break into consumer drones around the world.[84] GPS spoofing is a particular risk – whereby a hacker reprograms the GPS of a drone to make the drone believe it is somewhere else. According to roboticist Noel Sharkey, "it's very dangerous – if a drone is being directed somewhere using its GPS, [a spoofer] can make it think it's somewhere else and make it crash into a building, or crash somewhere else, or just steal it and fill it with explosives and direct somewhere."[85]

Insurgents in Iraq reportedly hacked into MQ-9 Reaper drones piloted by the US in 2008 and 2009.[86] If insurgents in Iraq could hack into the feed of the most advanced drone in the world, military robotics could be even more vulnerable when the hackers are advanced militaries.

As long as these concerns exist, some countries may severely restrict the mission set of military robotics. Just as in other domains, militaries are naturally attempting to bolster information security to ensure complete control, and one question is how much "risk" militaries will be willing to take when it comes to robotics.

Even if there were a way to ensure information security, bandwidth requirements could impose limitations on effectiveness (although some inhabited systems may share this issue as well). The growing dependence of the US, along with other advanced militaries, on information systems is well known. US power projection relies on high-bandwidth, networked forces. Specifically, secure access to an array of satellites circling the globe not only makes modern munitions precise, it is also critical for navigation and communications.[87]

What would happen, therefore, in a conflict where the US or another country dependent on satellite access lost access to them, either through jamming or adversary efforts to shoot down those satellites?[88] Attempts to develop nonsatellite means of connectivity aside, the challenge of operating without heavily networked forces raises important questions for military robotics. Severely degraded or denied access to

[84] Sternstein (2015); Tippenhauer et al. (2011). [85] Sydiongco (2012).
[86] MacAskill (2009). [87] Harrison (2013); Klausner (2002); Norris (2015).
[88] Krepinevich (2012). Others argue that China's military is also dependent on access to global information networks, meaning it would not attempt to disrupt satellites in a potential contingency.

information systems could render many military robotic systems irrelevant, because operators would lose access to those systems. The possibility of losing access to critical systems logically leads to doubt about how much militaries should rely on those systems.

Can Machine Autonomy Address Military Robotics Trade-offs?

How might militaries deal with the risk of hacking and bandwidth requirements? Addressing them may generate a new series of trade-offs. In contrast to current-generation systems, which often operate similarly to manned platforms, just without a pilot in the cockpit or driver's seat, another option is to delegate more decision making to the machine itself – to increase its level of autonomy.

In the context of this chapter, autonomy refers to the ability of a machine to complete a task, once activated, without further human intervention. Thus, autonomy does not necessarily imply anything about artificial intelligence, per se, and is best considered a continuum. As Horowitz and Scharre write, "it is meaningless to refer to a machine as 'autonomous' or 'semi-autonomous' without specifying the task or function being automated. Different decisions have different levels of complexity and risk."[89] It is essential to distinguish between incorporating autonomous features such as autopilot into military systems, and making the use of force autonomous.

The distinction between a lethal autonomous weapon system (LAWS) and a human-supervised autonomous system or semiautonomous system is that a person does not specify the targets.[90] Once activated, LAWS can target and fire without further human intervention. Even with weapon systems that some might consider an autonomous weapon, there could be significant human judgment and control involved in the use of force through the decision to activate the system. For LAWS that are munitions, for example, the use of force would be much more akin to the use of heat-seeking missiles today than something fundamentally new. Additionally, if a military commander under attack launches a swarm designed to destroy attacking aircraft and surface ships, it is not clear that the level of human control

[89] Scharre and Horowitz (2015a, 7); Heyns (2013).
[90] Scharre and Horowitz (2015a, 16).

or judgment over the use of force is fundamentally different from that of present systems.

For some of the issues above, autonomy could seem a potential solution, meaning militaries could theoretically be willing to trade off reduced human control to alleviate some of the vulnerability issues with remotely piloted military robotics. For example, midcourse hacking becomes possible because of the data link between military personnel and the platform. Preprogrammed systems, such as Tomahawk missile variants that only have internal maps and targeting computers that do not rely on GPS connectivity, cannot be hacked in flight because there is nothing to hack in-flight.

Moreover, while hacking or spoofing might still be a concern to some extent if military personnel can monitor a system or have an override switch, because there will still be a hackable data link, it could be possible to program a system such that, if it detects attempts at hacking or spoofing, it shuts off its external data connections and relies on a preprogrammed mission plan.

Autonomy could similarly provide advantages in environments with low or no data connectivity. For example, communication in the undersea environment is notoriously difficult. Underwater autonomous vehicles with long endurance could have the capacity to conduct long-term surveillance missions – or even pre-position for strike missions, with little if any communication.[91]

Of course, a move toward autonomy would come with its own trade-offs. A shift toward autonomy, even if there is still meaningful human control and accountability, means less overall control over a given use of force, or at least rethinking what human control means.

Most fundamentally, the potentially superior reaction time and accuracy of an autonomous system could make them attractive in a few cases. There are limited examples of autonomous weapon systems today, though they mostly do not exist. At the munitions level, the Israeli Harpy is a loitering cruise missile that, upon launch, turns on its seeker and looks for a particular type of radar. When it finds that radar, it targets the radar and accelerates for the kill. The political risk, and thus trade-off, is that, once launched, the Harpy cannot distinguish between whether the radar is on top of a hospital or a school, or a legitimate military target. There are other weapon systems, such

[91] Brannen (2014).

as the Long Range Anti-Ship Missile being developed by the US Navy that have autonomous features, but are not autonomous weapons.

At the platform level, while LAWS often invoke images of hunter-killer drones autonomously seeking particular targets, simple human-supervised autonomous systems exist today. Over thirty countries have deployed point-defense weapons such as the Phalanx and C-RAM to defend ships and military bases from attack for over thirty years. These systems have human operators that track potential incoming threats, such as cruise missiles, and then deploy the defensive systems in response (in the case of the Phalanx, a rapid-fire machine gun, in the case of C-RAM, missiles). The systems have an automatic mode that a person can activate if the number of incoming threats exceeds the ability of a person to respond. In that case, a computer system targets and fires against each incoming threat. Defensive activities such as force protection are one clear area where countries might have continuing interests in deploying LAWS.[92]

There are no existing fully autonomous weapon systems designed to orchestrate military campaigns, nor do militaries appear to be developing any. Delegating more control to autonomous weapon systems could come with risk, however, even at the munition or platform level. Autonomous systems, whether driverless cars or a Roomba, work best in a controlled environment in which the programmer can account for as many contingencies as possible. Accidents and malfunctions already occur in human-operated military systems around the world. Incorporating greater autonomy into weapon systems could exacerbate the potential for accidents and/or malfunctions if there is insufficient understanding of when and how to use the system.

A combination of potential programming errors, insufficient training of human operators concerning usage scenarios for activating LAWS, uncertainty from a complicated strategic environment, or rapid interactions with adversary LAWS could have unpredictable consequences. The worst-case scenario would be a military analogue to the May 6, 2010, "flash crash" of the US stock market triggered by automated trading algorithms. In that scenario, one or more of the issues above could lead to a "flash war."[93]

Given that militaries generally do not want to build weapons they cannot control, and the limited state of the technology, for the

[92] Scharre and Horowitz (2015b). [93] Scharre (2016).

foreseeable future, countries will likely only design and deploy a system in operating environments with simple terrain features and where any intruders are legitimate military targets (such as the DMZ on the Korean Peninsula), or where the target is extremely limited (e.g., radars in the case of the Israeli-made Harpy). The conceptual simplicity of the air and maritime environment, relative to complicated ground terrain, could make LAWS more technically plausible sooner in those spaces. If the complexity of an A2/AD battlespace is a key reason why a country might show interest in developing autonomous systems, however, that conceptual simplicity may be overwhelmed by an extremely complicated battlespace.

These concerns are some of the reasons to expect many militaries to move slowly and cautiously toward the use of LAWS if they decide to invest in them at all – and potentially to constrain their usage to the most predictable of operating environments.[94]

Conclusion

The continuing spread of military robotics, along with the greater capabilities apparent in next-generation systems, makes understanding their relative effectiveness critical for scholars and the policy community. Up to this point, military robotics, generally concentrated in the air, have been typically deployed to avoid putting friendly forces at risk and to gain advantages from control of the system and persistent surveillance. These systems come with trade-offs, however. Current-generation military robotic systems are generally inferior to their inhabited counterparts when it comes to overall firepower. Due to slow speeds, a lack of stealth, and limited, if any, self-defense capabilities, they are also less mobile and survivable than inhabited counterparts. The next generation of systems in the air and naval realms, however, is likely to narrow this capability gap.

Key limitations on current semiautonomous military robotics could continue restricting their effectiveness – in particular, fears of hacking and communication bandwidth requirements. This could create incentives for at least some countries to pursue autonomy in weapon systems. Some growth in autonomous military systems is likely inevitable.

[94] This chapter does not engage ethical and moral debates about LAWS. That is also an important topic for future research.

From improved autopilot to more advanced precision guidance, more autonomy can bolster military effectiveness. This does not, however, mean that LAWS, where the machine targets and fires itself, are inevitable. Though potentially attractive as a response to cyber threats, by walling off advanced military systems from network access entirely, countries are likely to move very slowly when it comes to developing LAWS. If technological developments in the area of autonomous systems continue over the next generation, however, or accelerate, this question could become trickier.

This chapter also suggests several futures areas for research. One issue that will shape the consequences of military robotics is the organizational challenge they present for militaries. Large powers have struggled when it comes to integrating new technologies into their militaries for hundreds of years, if not longer. For leading military powers, the risks can often seem much higher in abandoning technologies proven to "work," while rising powers or smaller powers have less to lose from embracing not just technologies, but operational concepts, that can disrupt the established balance of power.[95] To the extent that next-generation drones, and even more advanced military robotics, are framed as a threat to pilots, drivers, and others in high-status military occupations, they will trigger organizational antibodies in many militaries.[96] Understanding the interaction between the proliferation of these technologies and the organizational interests of militaries is a key task moving forward.

Another issue for future research is how robotics impact the balance between labor and capital in modern militaries, as well as defense procurement decision making. Over the last generation, the US has perfected a force structure that relies on a small number of advanced, capital-intensive platforms such as aircraft carriers, fighters, and bombers. This capital-intensive defense production strategy may only work, though, if the relative gap in force employment (the combination of technology, human capital, and organizational practices) between the US and other advanced militaries around the world is large.

Developing new versions of platforms as robotic technology improves and integrating software for autonomous operations or other

[95] Horowitz (2010). [96] Horowitz (2014); Work and Brimley (2014).

improvements could drive militaries toward shorter production cycles and smaller equipment buys than in the last two generations. If Google and other companies are producing cutting-edge robotics that are relatively easily adaptable to military uses, it means countries will have to produce new military technologies faster and faster to stay ahead, qualitatively. At the limit, these trends could presage the rise of quantity, once again, as opposed to quality, dictating how the most successful militaries in the world generate force. Regardless of this trend in particular, the continuing spread of military robotics will raise key questions about military effectiveness, forcing militaries to make both short-term trade-offs and long-term bets as they configure their militaries for the middle part of the twenty-first century.

References

Ackerman, Spencer. 2014. 41 Men Targeted but 1,147 People Killed: US Drone Strikes – the Facts on the Ground. *The Guardian*, November 24. www.theguardian.com/us-news/2014/nov/24/-sp-us-drone-strikes-kill-1147.

Adams, Lorraine, and Ayesha Nasir. 2010. Inside the Mind of the Times Square Bomber. *The Observer*, September 18. www.theguardian.com/world/2010/sep/19/times-square-bomber.

Axe, David. 2011. Taiwan's False Drone Hope? *The Diplomat*, August 17. http://thediplomat.com/2011/08/taiwans-false-drone-hope/.

BBC News. 2013. China "Flies First Stealth Drone" – Reports. *BBC News*, November 22. www.bbc.com/news/world-asia-china-25033155.

Bender, Jeremey. 2015. China May Have Just Unveiled a New Stealth Drone. *Business Insider*, January 14. www.businessinsider.com/china-unveiled-a-possible-stealth-drone-2015-1.

Bergen, Peter L., and Emily Schneider. 2014. Hezbollah Armed Drone? Militants' New Weapon. *CNN*, September 22. www.cnn.com/2014/09/22/opinion/bergen-schneider-armed-drone-hezbollah/.

Bloomberg Business. 2012. Rise of the Robot: Just How Big Is the Robotics Market? www.bloomberg.com/news/videos/b/44358604-ffe8–4692-a9d9–0f6f459222662012.

Boyle, Michael J. 2015. The Race for Drones. *Orbis* 59(1): 76–94.

Brannen, Samuel J. 2014. *Sustaining the US Lead in Unmanned Systems: Military and Homeland Considerations through 2025*. Washington, DC: Center for Strategic and International Studies, International Security Program. http://csis.org/files/publication/140227_Brannen_UnmannedSystems_Web.pdf.

Carroll, James. 2014. Report: Agricultural Robot Market Expected to Hit $16.3 Billion by 2020. *Vision Systems Design*, February. www.vision-systems.com/articles/2014/02/report-agricultural-robot-market-expected-to-hit-16–3-billion-by-20.html.

Carus, W. Seth. 1992. *Cruise Missile Proliferation in the 1990s*. Westport, CT: Center for Strategic and International Studies/Praeger.

Cenciotti, David. 2013. US F-22 Stealth Fighter Pilot Taunted Iranian F-4 Phantom Combat Planes over the Persian Gulf. *The Aviationist*, September 19. http://theaviationist.com/2013/09/19/f-22-f-4-intercept/.

Chase, Michael S., Kristen Gunness, Lyle J. Morris, Samuel K. Berkowitz, and Benjamin Purser. 2015. Emerging Trends in China's Development of Unmanned Systems. Research report, RAND Corporation. www.rand.org/content/dam/rand/pubs/research_reports/RR900/RR90/RAND_RR90.pdf.

Cillizza, Chris. 2013. The American Public Loves Drones. *The Fix* (blog), February 6. www.washingtonpost.com/blogs/the-fix/wp/2013/02/06/the-american-public-loves-drones/.

Cohen, Gili. 2015. Why Does Israel Need Three Different Missile Defense Systems? April 2. www.haaretz.com/news/diplomacy-defense/.premium-1.650293.

Cohen, Zachary. 2015. The F-35: Is the World's Most Expensive Weapons Program Worth It? *CNN*, July 16. www.cnn.com/2015/07/16/politics/f-35-jsf-operational-costs/.

Copp, Tara. 2015. The Air Force Wants a Futuristic Swarm of Exploding Autonomous Warbirds. *Washington Examiner*, February 9. www.washingtonexaminer.com/the-air-force-wants-a-futuristic-swarm-of-exploding-autonomous-warbirds/article/2559986.

Cronin, Audrey K. 2013. Why Drones Fail. *Foreign Affairs* 92(July/August): 44–54.

Davenport, Christian. 2007. State Repression and the Tyrannical Peace. *Journal of Peace Research* 44(4): 485–504.

Davis, Lynn E., Michael J. McNemey, James Chow, Thomas Hamilton, Sarah Harting, and Daniel Byman. 2014. Armed and Dangerous? UAVs and US Security. Research report, RAND Corporation. www.rand.org/content/dam/rand/pubs/research_reports/RR400/RR49/RAND_RR49.pdf.

Defense Science Board. 2012. *Task Force Report: The Role of Autonomy in DoD Systems*. Washington, DC: Office of the Undersecretary of Defense for Acquisition, Technology and Logistics. http://fas.org/irp/agency/dod/dsb/autonomy.pdf.

Defense Security Cooperation Agency. 2015. The Netherlands – MQ-9 Reapers. www.dsca.mil/sites/default/files/mas/netherlands_14–55.pdf.

Defense Update. 2013. China's Lijian UCAV "Sharp Sword" Spotted Taxiing. May 18. http://defense-update.com/20130518_lijian-sharp-sword-taxi.html.

Eaglen, Mackenzie, and Bryan McGrath. 2015. Future Warfare: The X-47b. *Real Clear Defense*, April 25. www.realcleardefense.com/articles/2015/04/25/future_warfare_the_x-47b_107890.html.

Economist. 2013. Speed Is the New Stealth. June 1. www.economist.com/news/technology-quarterly/21578522-hypersonic-weapons-building-vehicles-fly-five-times-speed-sound.

Everstine, Brian. 2014. Why the Air Force Wants to Keep Global Hawks and Retire U-2s. *Air Force Times*, March 5. http://archive.airforcetimes.com/article/20140305/NEWS04/303050029/Why-Air-Force-wants-keep-Global-Hawks-retire-U-2s.

Farley, Robert. 2014. After China: The Proliferation of Cruise Missiles. *The Diplomat*, July 3. http://thediplomat.com/2014/07/after-china-the-proliferation-of-cruise-missiles/.

Federation of American Scientists. 2000. F/A-18 Hornet. April 25. http://fas.org/man/dod-101/sys/ac/f-18.htm.

Freedberg, Sydney J., Jr. 2015. DepSecDef Explores New Missile Defense Approach. *BreakingDefense*, August 11. http://breakingdefense.com/2015/08/depsecdef-launches-new-missile-defense-approach/.

Fuhrmann, Matthew, and Michael C. Horowitz. 2017. Droning On: Explaining the Proliferation of Unmanned Aerial Vehicles. *International Organization* 71(2): 397–418.

General Accounting Office. 2000. Comprehensive Strategy Needed to Improve Ship Cruise Missile Defense. GAO/NSIAD-00-149. www.gao.gov/assets/230/229270.pdf.

Gertler, Jeremiah. 2014. F-35 Joint Strike Fighter (JSF) Program. www.fas.org/sgp/crs/weapons/RL30563.pdf.

Gilli, Andrea, and Mauro Gilli. 2016. The Diffusion of Drone Warfare? Industrial, Organizational and Infrastructural Constraints: Military Innovations and the Ecosystem Challenge. *Security Studies* 25(1): 50–84.

Global Industry Analysts. 2012. Global Defense Robotics Market to Reach US$7.5 Billion by 2018. *PRWeb*, August 8. www.prweb.com/releases/defense_robotics/military_robots/prweb9780170.htm.

Goldman, Emily O., and Leslie C. Eliason. 2003. Introduction: Theoretical and Comparative Perspectives on Innovation and Diffusion. In Emily O. Goldman and Leslie C. Eliason, eds., *The Diffusion of Military Technology and Ideas*, 1–30. Stanford, CA: Stanford University Press.

Hardy, James. 2014. Japan Announces Global Hawk, Osprey, Hawkeye Purchases. *Jane's Defence Weekly*, November 20. www.janes.com/article/46125/japan-announces-global-hawk-osprey-hawkeye-purchases.

Harrison, Todd. 2013. The Future of Milsatcom. http://csbaonline.org/publications/2013/07/the-future-of-milsatcom/.

Hen, Ng Eng. 2014. Speech at Committee of Supply Debate 2014. www.mindef.gov.sg/imindef/press_room/official_releases/sp/2014/06mar14_speech.html.

Heyns, Christof. 2013. Report of the Special Rapporteur on Extrajudicial, Summary or Arbitrary Executions. www.ohchr.org/Documents/HRBodies/HRCouncil/RegularSession/Session23/A-HRC-23–47_en.pdf.

Horowitz, Michael C. 2010. *The Diffusion of Military Power: Causes and Consequences for International Politics*. Princeton, NJ: Princeton University Press.

2014. The Looming Robotics Gap: Why America's Global Dominance in Military Technology Is Starting to Crumble. *Foreign Policy Magazine*, May/June. http://foreignpolicy.com/2014/05/05/the-looming-robotics-gap/.

2016. Public Opinion and the Politics of the Killer Robots Debate. *Research & Politics* 3(1): 1–8.

Horowitz, Michael C., Sarah Kreps, and Matthew Fuhrmann. 2016. The Consequences of Drone Proliferation: Separating Fact from Fiction. *International Security* 41(2): 7–42.

Horowitz, Michael, and Dan Reiter. 2001. When Does Aerial Bombing Work? Quantitative Empirical Tests, 1917–1999. *Journal of Conflict Resolution* 45(April): 147–173.

Jervis, Robert. 1978. Deterrence Theory Revisited. Acis Working Paper 14, Center for Arms Control and International Security University of California, Los Angeles, CA.

Johnston, Patrick B., and Anoop K. Sarbahi. Forthcoming. The Impact of US Drone Strikes on Terrorism in Pakistan. *International Studies Quarterly*.

Jones, Bruce. 2014. Russia to Build Unmanned Airships for Arctic Military Patrols. *Jane's Defence Weekly*, March 13. www.janes.com/article/35324/russia-to-build-unmanned-airships-for-arctic-military-patrols.

Karp, Aaron. 1996. *Ballistic Missile Proliferation: The Politics and Technics*. Oxford: Oxford University Press.

Kasper, Joakim. 2015a. AGM-114 Hellfire Missile. *Aeroweb*, April 8. www.bga-aeroweb.com/Defense/AGM-114-Hellfire-Missile-System.html.

2015b. Joint Direct Attack Munition. *Aeroweb*, July 15. www.bga-aeroweb.com/Defense/JDAM.html.

Kington, Tom, and Pierre Tran. 2015. New Effort at European UAV Raises Hopes. *Defense News*, May 23. www.defensenews.com/story/defense/air-space/isr/2015/05/23/france-germany-italy-uav-male-design-study/27545807/.

Klausner, Kurt A. 2002. Command and Control of Air and Space Forces Requires Significant Attention to Bandwidth. *Air and Space Power Journal* 16(4): 69–77.

Krepinevich, Andrew F. 2012. Why Airsea Battle. http://csbaonline.org/publications/2010/02/why-airsea-battle/.

Kreps, Sarah. 2014. Flying under the Radar: A Study of Public Attitudes towards Unmanned Aerial Vehicles. *Research and Politics* 1(1): 1–7.

LaGrone, Sam. 2015. Navy Conducts Successful Test of Aerial Refueling with X-47b, UCAS-D Program Ending. *USNI News*, April 22. http://news.usni.org/2015/04/22/navy-conducts-successful-test-of-aerial-refueling-on-x-47b-ucas-d-program-ending.

Lewis, Michael W. 2013. Drones: Actually the Most Humane Form of Warfare Ever. *The Atlantic*, August 21. www.theatlantic.com/international/archive/2013/08/drones-actually-the-most-humane-form-of-warfare-ever/278746/.

Lieber, Keir A. 2000. Grasping the Technological Peace: The Offense-Defense Balance and International Security. *International Security* 25(1): 71–104.

Litovkin, Dmitry. 2014. Russia to Bring First Unmanned Combat Aircraft into Service by 2020. *Russia Beyond the Headlines*, June 22. http://rbth.com/defence/2014/06/22/russia_to_bring_first_unmanned_combat_aircraft_into_service_by_20_37609.html.

Lockheed Martin. 2015. U-2 Specifications. www.lockheedmartin.com/us/products/u2/u-2-specifications.html.

MacAskill, Ewan. 2009. US Drones Hacked by Iraqi Insurgents. *The Guardian*, December 19.

Mahnken, Thomas G. 2005. The Cruise Missile Challenge. www.csbaonline.org/wp-content/uploads/2011/03/05.03.10-Cruise-Missile.pdf.

Mettler, Simon A., and Dan Reiter. 2013. Ballistic Missiles and International Conflict. *Journal of Conflict Resolution* 57(5): 854–880.

Michel, Christopher. N.d. U-2s Still Flying High. www.usni.org/print/6705.

Miller, Christopher. 2015. Drones Help Save Migrants in the Mediterranean. *Mashable*, August 21. http://mashable.com/2015/08/21/drones-help-find-migrants-in-the-mediterranean-sea/.

Missile Technology Control Regime. 1987. Frequently Asked Questions (FAQs). www.mtcr.info/english/FAQ-E.html.

Norris, Pat. 2015. Satellite Programs in the United States. In K. U. Schrogl, P. L. Hays, J. Robinson, D. Moura, and C. Giannopapa, eds., *Handbook of Space Security*, 743–774. New York: Springer.

Northrop Grumman. 2012. RQ-4 Block 40 Global Hawk. www.northropgrumman.com/Capabilities/GlobalHawk/Documents/Datasheet_GH_Block_40.pdf.

Office of the Secretary of Defense. 2014. Annual Report to Congress: Military and Security Developments Involving the People's Republic of China. www.defense.gov/pubs/2014_DoD_China_Report.pdf.

Orpaz, Inbal. 2013. Israel Navy's First Unmanned Surface Vehicle Keeps an Eye on the Sea. *Haaretz*, April 17. www.haaretz.com/news/diplomacy-defense/israel-navy-s-first-unmanned-surface-vehicle-keeps-an-eye-on-the-sea.premium-1.515978.

Osborn, Kris. 2015. Air Force's New Unmanned Strategy Has F-35 Pilots Flying Drones. *Military.com*, May 19. www.military.com/daily-news/2015/05/19/air-forces-new-unmanned-strategy-has-f35-pilots-flying-drones.html.

Peck, Michael. 2015. How the US Army Plans to Defeat the Unthinkable: Drone Swarms. *The National Interest*, January 18. http://nationalinterest.org/feature/how-the-us-army-plans-defeat-the-unthinkable-drone-swarms-12057.

Prigg, Mark. 2014. Who Goes There? Samsung Unveils Robot Sentry That Can Kill from Two Miles Away. *Daily Mail*, September 15. www.dailymail.co.uk/sciencetech/article-2756847/Who-goes-Samsung-reveals-robot-sentry-set-eye-North-Korea.html.

Robson, Seth. 2014. Air Force Plans Drone Upgrade to Replace U-2 Planes. *Stars and Stripes*, March 15. www.stripes.com/news/air-force-plans-drone-upgrade-to-replace-u-2-planes-1.272289.

Scharre, Paul. 2016. Autonomous Weapons and Operational Risk. Working paper, Center for a New American Security. https://s3.amazonaws.com/files.cnas.org/documents/CNAS_Autonomous-weapons-operational-risk.pdf.

Scharre, Paul, and Michael C. Horowitz. 2015a. An Introduction to Autonomy in Weapon Systems. Working paper, Center for a New American Security. www.cnas.org/sites/default/files/publications-pdf/Ethical%20Autonomy%20Working%20Paper_021015_v02.pdf.

2015b. Keeping Killer Robots on a Tight Leash. *Defense One*, April 14. www.defenseone.com/ideas/2015/04/keeping-killer-robots-tight-leash/110164/.

Smalley, David. 2014. The Future Is Now: Navy's Autonomous Swarmboats Can Overwhelm Adversaries. www.onr.navy.mil/Media-Center/Press-Releases/2014/autonomous-swarm-boat-unmanned-caracas.aspx.

Soper, Karl. 2015. Russia Formulates UAV Development Programme. *Jane's Defence Weekly*, May 28. www.janes.com/article/51652/russia-formulates-uav-development-programme.

Sternstein, Aliya. 2015. How to Hack a Military Drone. *Defense One*, April 29. www.defenseone.com/technology/2015/04/how-hack-military-drone/111391/.

Svolik, Milan W. 2013. Contracting on Violence: The Moral Hazard in Authoritarian Repression and Military Intervention in Politics. *Journal of Conflict Resolution* 57(5): 765–794.

Sydiongco, David. 2012. Research Team Hacks Surveillance Drone with Less Than $1,000 in Equipment. *Slate*, July 2. www.slate.com/blogs/future_tense/2012/07/02/hacked_surveillance_drone_with_spoofed_gps_system_demonstrates_uav_security_flaws_.html.

Tadjdeh, Yasmin. 2014. The US Navy Is Working on an Incredible Next-Generation Underwater Drone. *National Defense Magazine*, August 20. www.businessinsider.com/the-navys-new-underwater-drone-2014-8.

Talmadge, Caitlin. 2015. *The Dictator's Army: Battlefield Effectiveness in Authoritarian Regimes*. Ithaca, NY: Cornell University Press.

Tatlow, Didi K. 2014. China Said to Deploy Drones after Unrest in Xinjiang. *New York Times*, August 19. http://sinosphere.blogs.nytimes.com/2014/08/19/china-said-to-deploy-drones-after-unrest-in-xinjiang.

Teal Group Corporation. 2014. Teal Group Predicts Worldwide UAV Market Will Total $91 Billion in Its 2014 UAV Market Profile and Forecast. www.tealgroup.com/index.php/about-teal-group-corporation/press-releases/118–2014-uav-press-release.

Tippenhauer, Nils Ole, Christina Pöpper, Kasper Bonne Rasmussen, and Srdjan Capkun. 2011. On the Requirements for Successful GPS Spoofing Attacks. Paper presented at the 18th ACM conference on Computer and Communications Security.

Transparency Market Research. 2014. Mobile Robotics Market Worth US$14,202.2 Million by 2019. *Global Newswire*, October 22. http://globenewswire.com/news-release/2014/10/22/675241/10103825/en/Mobile-Robotics-Market-Worth-US-14–202–2-million-by-2019-Transparency-Market-Research.html.

US Air Force. 2015. MQ-9 Reaper. www.af.mil/AboutUs/FactSheets/Display/tabid/224/Article/104470/mq-9-reaper.aspx.

US Navy. 2009. F/A-18 Hornet Strike Fighter. www.navy.mil/navydata/fact_display.asp?cid=1100&tid=200&ct=1.

2014. Tomahawk Cruise Missile. www.navy.mil/navydata/fact_display.asp?cid=2200&tid=1300&ct=2.

Van Evera, Stephen. 1999. *Causes of War*. Ithaca, NY: Cornell University Press.

Villarejo, Esteban. 2015. Spain Eyes Offer of Reaper, Heron. *Defense News*, September 11. www.defensenews.com/story/defense/air-space/isr/2015/09/11/spain-eyes-offer-reaper-heron/72057400/.

Wheeler, Winslow. 2012. The MQ-9's Cost and Performance. *Time*, February 28. http://nation.time.com/2012/02/28/2-the-mq-9s-cost-and-performance/.

Wong, Kelvin. 2014. Russia to Deploy Armed UGVs at Ballistic Missile Bases. *International Defense Review*, March.

Work, Robert O. 2015. Speech at CNAS Defense Forum. December 14. www.defense.gov/News/Speeches/Speech-View/Article/634214/cnas-defense-forum.

Work, Robert O., and Shawn Brimley. 2014. *20yy: Preparing for War in the Robotic Age*. Washington, DC: Center for a New American Security. www.cnas.org/sites/default/files/publications-pdf/CNAS_20YY_WorkBrimley.pdf.

Zegart, Amy. 2016. Coercion in a Drone World. Unpublished manuscript, Stanford University.

Zenko, Micah. 2014. America's 500th Drone Strike. *The Atlantic*, November 21. www.theatlantic.com/international/archive/2014/11/america-just-launched-its-500th-drone-strike/383039/.

Zenko, Micah, and Sarah Kreps. 2014a. Limiting Armed Drone Proliferation. Special press report 69, Council on Foreign Relations. www.cfr.org/drones/limiting-armed-drone-proliferation/p33127.

2014b. The Next Drone Wars: Preparing for Proliferation. *Foreign Affairs* 93(2): 68–79.

7 *Too Much of a Good Thing?*

Conventional Military Effectiveness and the Dangers of Nuclear Escalation

CAITLIN TALMADGE

Escalation control is an often overlooked component of military effectiveness, one that brings into sharp relief the security-related trade-offs states must manage in their efforts to fight and win wars. As the introduction noted, effective militaries must meet the goals set for them by political leaders at an acceptable level of cost. As wars escalate, so do costs – in economic, diplomatic, and human terms. As a result, managing escalation to keep the war within politically acceptable bounds along all these dimensions is fundamental to military effectiveness. Indeed, efforts to achieve military effectiveness that provoke adversary escalation and the rise of new security threats present a particularly vexing trade-off that states must carefully manage.

Nowhere is this imperative more consequential than in preventing nuclear escalation – that is, the actual use of nuclear weapons – in conventional wars between nuclear-armed states. A large and useful literature has explored the requirements for conventional military success, as noted in the introduction. Yet perhaps because this literature blossomed during the long post–Cold War interval in which the prospect of conventional war between nuclear-armed states appeared to have receded, the general assumption has been that states should have the goal of maximizing conventional military effectiveness, and the debate has been about how to do so. Unfortunately, plausible scenarios for conventional conflict among nuclear-armed states – for example, Russia, China, the US, North Korea, India, and Pakistan – have returned to the forefront of the global security environment, and thus it is imperative to broaden the discussion of military effectiveness to include again considerations of nuclear escalation management.

As the well-developed but at times forgotten nuclear strategy literature has long noted, conventional military effectiveness can heighten the probability of adversary nuclear escalation. Not only would opponent nuclear escalation under such circumstances likely prevent military victory in purely operational terms for the side subject to the attack, but nuclear use also would impose such high costs that an operational victory would be politically meaningless. After all, no state would describe its military strategy as effective if the strategy pushed the opponent to resort to nuclear weapons. Hence in conventional contests between nuclear-armed states, military effectiveness is not solely about executing operational and tactical tasks. It is also about calibrating the use of military force such that the opponent capitulates rather than turns to nuclear weapons.

The chapter develops and explores the implications of this argument in three parts. The first part defines escalation control in relation to broader debates about trade-offs in the pursuit of conventional effectiveness. It then explains where nuclear escalation fits into this framework and draws on the nuclear strategy literature to theorize about two pathways by which a state's conventional military effectiveness could heighten the risk of opponent nuclear escalation. The first is nuclear escalation to stave off conventional defeat. The second is nuclear escalation in response to what I call "conventional counterforce," that is, an effort by one state to use conventional military force to destroy or degrade an opponent's nuclear arsenal or ability to use that arsenal, for example, by using conventional forces to degrade an opponent's nuclear-relevant command-and-control assets.

The second and third parts of the chapter then examine these two potential nuclear escalatory pathways in more depth, assessing for each what specific forms escalation might take given the differing motives and evaluating the likelihood that future conventional wars might stimulate such motives. Given the (fortunate) dearth of nuclear wars, the empirical focus is on the behavior of nuclear- or WMD-armed states in past conventional conflicts: the Korean War, the Arab–Israeli Wars of 1967 and 1973, India and Pakistan from 1990 to 2008, and Iraq in 1991. The chapter also draws on known features of nuclear states' past and present postures and doctrines in an effort to ascertain how states might react to nuclear escalatory pressures in the future, particularly in scenarios involving North Korea, Russia, China, or the US.

The chapter concludes that although nuclear escalation is still a very low-probability event, in some situations conventional military effectiveness actually could heighten the likelihood of it occurring. In fact, states at times may actively attempt to manipulate this risk. The danger thus represents a serious trade-off in the design of conventional military campaigns, one sometimes overlooked in standard assessments of military effectiveness and that holds important implications for scholarship and policy.

Nuclear Escalation Control as a Component of Military Effectiveness

The notion that all good things do not go together is a recurring theme in the study of military effectiveness. Numerous wars bear out the notion that maximizing conventional advantages, especially at the tactical and operational levels of war, can create other security risks. For example, by many definitions, the US military was operationally and tactically successful in its battles against communist forces in South Vietnam, but these "successes" at killing insurgents were ultimately counterproductive.[1] They rallied peasants to resist rather than support the government in Saigon.[2] The robust US military role in these battles also decreased the incentives for South Vietnamese leaders to improve their own military forces.[3] In short, not only did the US military fail to achieve its strategic and political goals in the war, but in a very real sense this failure occurred because of – not in spite of – the military's tactical and operational successes.

Surprisingly, however, the subject of escalation control is missing from much of the effectiveness literature.[4] Defined simply, escalation is "an increase in the intensity or scope of conflict that crosses threshold(s) considered significant by one or more of the participants."[5] In practice, this increase could take many forms: a raising of the stakes involved in conflict, an expansion of the number of war participants, the lengthening of a war's duration, or an intensification of the level of violence employed by one or both sides.[6]

Escalation can occur both within and across the various levels of war. For example, the US decision to use airpower to support Iraqi security forces attempting to re-take the city of Tikrit from Islamic State

[1] Summers (2009). [2] Krepinevich (1988). [3] Talmadge (2015).

[4] Tellingly, one of the best general works on escalation is more than forty years old. See Smoke (1977).

[5] Morgan et al. (2008).

[6] See Smoke (1977, Chapter 2); Morgan et al. (2008). On the expansion of war participants, see Shirkey (2009). On the challenges of war termination, see Reiter (2009).

militants in spring 2015 constituted tactical escalation. But the operation also had significant implications for the political effectiveness of the US use of force, because it ran the risk of protecting or even empowering Iranian-backed Shia militias also aligned with the Iraqi forces, which in turn was likely to decrease support from the US's Arab partners.[7] Similarly, the German decision to adopt the Schlieffen Plan in 1914 constituted an operational escalation, but it also had strategic and political implications because the incursion into Belgium provoked British entry into the war.[8] As these examples suggest, a truly effective military has to be able to achieve tactical and operational successes but ensure that such gains do not provoke allied or adversary reactions that impose disproportionate strategic or political costs.[9]

Escalation control in conventional wars between nuclear-armed states presents a potential variant of this broader conundrum, albeit with a twist. In the type of purely conventional contexts just discussed, if the stakes are high enough a state can decide to pursue all-out military victory through the application of brute force at the tactical and operational levels, accepting whatever strategic or political repercussions may ensue. Under conditions of mutually assured destruction, however, the brute force conventional strategy is arguably off the table, because it can result in adversary nuclear escalation that would nullify all traditional notions of military victory. As Jervis has put it, "No one can win an all-out nuclear war."[10] Thus, in a world of secure second strike forces, states may face not so much trade-offs between different types of effectiveness across the different levels of warfare, as they do in the conventional context, but rather trade-offs between military effectiveness and the risk of nuclear annihilation.

That said, the intensity of this trade-off hinges in significant part on whether there are any circumstances in which rational states would use nuclear weapons at all in a situation of mutual vulnerability. At first the answer might seem to be no. As Jervis has observed, "A rational strategy for the employment of nuclear weapons is a contradiction in terms."[11] Threats of all-out nuclear use are suicidal against an adversary who can retaliate and therefore lack credibility. In fact, the

[7] *New York Times* (2015).
[8] Kissinger (1995, Chapters 7 and 8); Snyder (1984, Chapters 4 and 5).
[9] In fact, adversaries may tacitly collude to keep military operations secret in order to avoid these sorts of escalatory pressures. Carson (2016).
[10] Jervis (1989, 1–8). [11] Jervis (1984, 19).

"stability–instability paradox" suggests that such threats are so incredible that a condition of mutual strategic nuclear vulnerability can actually allow for "safe" lower-level conflict, even up to the point of tactical nuclear use.[12]

Still, the firebreak between conventional and nuclear war under conditions of mutual vulnerability may not be as strong as the stability–instability paradox would imply, for two reasons. First, the brinkmanship literature suggests that precisely because states recognize that threats to use nuclear weapons deliberately are incredible, states will engage in competitions in risk taking that invoke Schelling's famous "threat that leaves something to chance."[13] In other words, states will not directly threaten to use nuclear weapons in an all-out attack on the adversary but will bid up what Powell describes as the "autonomous risk" that such escalation could occur through a process over which both sides lose control, such as a mistaken launch on warning or unauthorized use of nuclear weapons. Nuclear crises are therefore contests over which side can tolerate more of this risk before blinking, but there is always a chance that because of incomplete information someone blinks too late and general nuclear war occurs.[14]

Second, states can engage in rational forms of limited nuclear escalation, either for coercive purposes (to signal resolve to the other side in a bargaining process) or for military purposes (to achieve a rapid tactical or operational advantage that would be difficult or impossible to achieve with conventional forces). Rational states plausibly could believe that a limited initial use of nuclear weapons for these purposes would not lead to all-out nuclear war.[15] Indeed, the history of US nuclear strategy in the Cold War is largely the story of an attempt to develop a credible threat of limited nuclear use even under the condition of mutual vulnerability with the Soviets.[16]

[12] The term comes from Snyder (1965), although analysts invoking the term often differ in their definitions of the paradox and especially the term *stability*. See also Jervis (1979–1980, 619).

[13] Schelling (1980, 187). [14] Powell (1987, 719). See also Powell (1985).

[15] Powell (1985); Jervis (1979–1980, 628–629).

[16] Colby (2014a); Gavin (2012). This may be why the superpowers continued to pursue nuclear superiority even under conditions of mutually assured destruction – because of a belief that superiority afforded bargaining advantages. See Jervis (1979–1980); Kroenig (2013); Sechser and Fuhrmann (2013).

Furthermore, an opponent's conventional military effectiveness could heighten two specific motives for a state to engage in this sort of limited nuclear use, meaning that there could be a direct trade-off between the opponent's pursuit of conventional military effectiveness and its desire to avoid provoking nuclear escalation.[17] First, a state could escalate to the use of nuclear weapons in order to stave off conventional defeat, especially if such defeat threatened significant loss of home territory. In such a scenario, a state fearing conventional defeat could use nuclear weapons to target the military forces, bases, and weapons of its conventionally stronger opponent in an effort to halt that opponent's conventional military operations. This sort of nuclear escalation could lead to such rapid and massive losses for the conventionally stronger side that it might find it difficult if not impossible to achieve military victory with reference to the original campaign goals, even in a purely operational sense and even if the conventionally stronger side engaged in nuclear retaliation. In short, a campaign that initially was so conventionally effective that it prompted the opponent to resort to nuclear weapons ultimately would not be effective at all.

Second, a state could escalate to nuclear use in a situation where it believed that its opponent's conventional military campaign sought to destroy the state's nuclear deterrent. This could be distinct from the first scenario in which the state believed that it was simply losing a high-stakes conventional war and that nuclear escalation offered a hail Mary pass at avoiding certain defeat. Even if a state were not about to lose an ongoing conventional war, it could believe that its opponent was using the conventional war to degrade the state's nuclear capabilities. I call this scenario "conventional counterforce," that is, an effort by one state to use conventional military force to destroy or degrade an opponent's nuclear arsenal or ability to use that arsenal, for example, by using conventional forces to degrade an opponent's nuclear-relevant command-and-control assets or to destroy conventional platforms or forces relevant to the opponent's operation of nuclear weapons. The target of such a campaign, or perceived campaign, might believe it faced the impending loss of its nuclear arsenal and decide to escalate to nuclear use while it still had the ability to do so, in ways I outline more fully in the next section.

[17] Powell (2015, 591) refers to this as the "trade-off between power and risk."

The key point is that, as with the first type of nuclear escalation, the resulting nuclear attacks against military targets might occur with such speed and destructive force as to make victory quite difficult for the other side to achieve in a purely operational sense. In addition, attacks against other targets could raise the human, economic, and environmental costs of the war to such a high level that "victory" for either side would have little meaning. Both scenarios suggest that there could indeed be intense effectiveness trade-offs in the design of conventional campaigns by and against nuclear-armed states. The next two sections elaborate upon and evaluate these scenarios in turn.

Nuclear Escalation to Prevent Conventional Defeat

States often acquire nuclear weapons because of a concern about the prospect of being overrun in a conventional military attack on their home territory. Naturally, then, states also may adopt doctrines that explicitly threaten first use of nuclear weapons in the event that conventional defeat or territorial dismemberment seem imminent.[18] The goal of such a posture would not be nuclear preemption or counterforce, that is, an effort to destroy the opponent's nuclear force.[19] Rather, the goal would be to staunch further territorial losses once a conventional war had begun. Limited nuclear use could do this by militarily stalemating an otherwise unstoppable conventional attack, or by helping the state generate coercive bargaining advantages against a less resolved but conventionally superior opponent, or both.[20]

States that share a land border with a conventionally superior opponent face strong pressures to adopt this approach. Notably, there is nothing accidental, inadvertent or unauthorized about this scenario for nuclear launch, though one can imagine such a posture increasing the risks of these dangers.[21] Brandishing or using nuclear weapons certainly is not ideal, but states may conclude that they have little alternative if an adversary is about to overrun the homeland. In

[18] For more on the logic of such a posture and possible variants, see Castillo (2014). States also can create this risk through alliances that lead patrons to station nuclear weapons on their territory. See Reiter (2014).

[19] For background on these aspects of nuclear strategy, see Glaser (1990); Jervis (1984, 1989); Long and Green (2014); Green and Long (2015).

[20] Powell (1987, 2015).

[21] Blair (1985, 1993); Sagan (1995); Carter et al. (1987).

fact, the more conventionally effective a given adversary is, the more likely it is that a state will feel pressure to adopt this sort of escalatory posture, especially if it has suffered territorial conquest or conventional defeat in living memory.

Such an attack could involve the strategic use of nuclear weapons against an adversary's homeland, as a form of punishment, or the tactical use of nuclear weapons against an adversary's forces in the field, as a form of denial, or a combination of the two.[22] Either type of use would clearly pose a serious challenge to the military effectiveness of the target state, either by imposing unbearable human, economic, and political costs, or by preventing victory in a purely operational sense due to the destruction of the target state's military capacity. Hence, it is a clear example of a trade-off that states face in deciding how to use and operate their conventional military forces.

Certainly, the experience of the Cold War testifies to the ways in which one side's conventional military prowess can actually drive the other side to adopt a deliberately escalatory nuclear posture in response. For example, the fear of being rapidly overrun in a conventional Warsaw Pact campaign was the primary motivation for NATO to threaten to use nuclear weapons first in the event of war – either tactically on the battlefield, or potentially in strategic attacks on Soviet territory. Had Warsaw Pact forces not had the perceived ability to so rapidly overrun key NATO members, it is unlikely that such a NATO doctrine would have been deemed worth the retaliatory risks it posed. France acquired its own nuclear arsenal in large part because of its concern about such dangers, and no doubt the French approach was born partly of the recent experience of having been conventionally conquered and occupied.[23]

Beyond the Cold War European context, other states also have responded to the danger of rapid conventional defeat with threats to use nuclear weapons first. For example, Israel's undeclared nuclear capabilities have long been understood to threaten this possibility – unsurprising given that the Arab conventional threat was the primary driver of Israeli acquisition of nuclear weapons in the first place.[24]

[22] On the distinction between punishment and denial, see Snyder (1961).

[23] On NATO and France, see Lieber and Press (2013a); Narang (2014, Chapter 6); Sagan (1989, 64, 77); Snyder (1965, 193).

[24] Cohen (1999).

Similarly, Pakistani leaders sought nuclear weapons after several devastating conventional defeats at the hands of India, most notably the loss of west Pakistan (now Bangladesh) in 1971. Pakistani nuclear doctrine has never disavowed the first use of nuclear weapons and instead emphasizes Pakistani willingness to employ such weapons first in the event that conventional military defeat looms in conflict with India.[25]

Notably unlike the Cold War, which stayed cold, both Israel and Pakistan have experienced since the acquisition of nuclear weapons conventional crises and conflicts that very nearly went hot. These episodes provide a useful opportunity to probe exactly how the posited trade-off between conventional military effectiveness and nuclear escalatory risk has played out empirically. In general, the Israeli and Pakistani experiences lend credence to the notion that an opponent's conventional military effectiveness can actually lead states to move toward nuclear escalation: both Israel and Pakistan have come uncomfortably close to using nuclear weapons in situations where their adversaries threatened conventional defeat.

In the 1967 Six-Day War, for example, Israel was prepared to use its nascent nuclear capability in the event that Arab armies threatened to overrun the country.[26] As Vipin Narang notes, "Israel – evidently unbeknownst to Egypt, the USSR, or the US – assembled several nuclear weapons for potential use during the conflict," possibly to help signal the need for US intervention on Israel's behalf.[27] Such use ended up being unnecessary because of Israel's stunningly successful preemptive attack against the Arab air forces, but Israel's preparations revealed the dangers of nuclear escalation that could have arisen had the Arab militaries proved to be more conventionally adroit.[28]

The 1973 Yom Kippur War did see more conventionally effective Arab military performance and with it a much heightened risk of Israeli nuclear use. The very fact that Egypt and Syria successfully launched a coordinated surprise attack on Israel, employing significantly upgraded conventional capabilities thanks to extensive training and Soviet weapons, shocked Israeli leaders and made them fear that the tiny Jewish state might soon be overrun entirely. Although Arab ambitions (especially Egypt's) were actually more limited – mainly to

[25] Ganguly and Kapur (2012). [26] Cohen (1999). [27] Narang (2014, 286).
[28] Narang (2014, 286). On the decision making surrounding the preemptive attack, see Reiter (1995, 16–19).

get Israel back to the negotiating table in order to re-claim territory lost in 1967 – Israeli leaders did not know this at the time.[29]

In fact, there is little evidence that Egypt and Syria successfully signaled otherwise.[30] In the run-up to the war, both Sadat and Assad spoke of "total confrontation" with Israel.[31] Within a few days of the war starting, Egyptian forces in the south looked as though they were poised to march across the Sinai and into Israel itself, while Syrian forces in the north were moving closer and closer to crossing the Jordan River and threatening the Israeli population. Israeli forces were running out of ammunition and equipment, and the two Arab armies might soon cut Israel in half.[32]

Caught flat-footed by what it perceived as a potentially existential threat, Israel turned to its nuclear weapons as an option for rectifying the conventional military situation. Israel conducted operational checks of its Jericho ballistic missiles in ways visible to US intelligence, likely to try pressure the US into a resupply but also simply so that the missiles would be ready to use if needed.[33] Such an occasion never arrived, although not because the Arab armies purposely limited their campaign in the face of possible Israeli nuclear use; it is unlikely that Egypt and Syria were even aware of the Israeli nuclear preparations. Rather, the Arab advances began to slow under the weight of heavy casualties, logistical constraints, declining cohesion, and other limitations arising from the conduct of their own campaign. Meanwhile, the US did resupply Israel, giving a strong boost to Israeli conventional effectiveness.[34] In short, the conventional threat to Israel receded, and with it the risk of nuclear escalation, but the case illustrates that the link between conventional threats and nuclear escalatory risks may be tighter than often recognized.

Indo-Pakistani crises and conflicts reflect a similar dynamic in which Indian conventional effectiveness has raised risks of Pakistani nuclear use. In the 1990 Kashmir Compound crisis, for example, Pakistan appears to have "assembled and moved nuclear weapons" in anticipation of a coming conventional fight with India.[35] During the 1999 Kargil War, the Pakistani military again appears to have begun to prepare nuclear weapons for launch – in this case apparently without the

[29] Colby et al. (2013, 13). [30] Colby et al. (2013, 13).
[31] Quoted in Narang (2014, 291). [32] Narang (2014, 289).
[33] Narang (2014, 289–292); Colby et al. (2013, 29–49).
[34] Colby et al. (2013, 48). [35] Narang (2014, 265).

knowledge of the civilian prime minister Nawaz Sharif. India became aware of the preparations and sharply limited its conventional operations in response.[36] The pattern repeated itself in the 2001–2002 crises surrounding India's Operation Parakram. In response to large-scale Indian preparations for conventional operations against Pakistan, Pakistan "test-fired three nuclear-capable ballistic missiles" and amped up nuclear threats in its public rhetoric. Indian policy makers again chose to limit and then eventually call off their conventional military operations.[37] A similar dynamic unfolded in 2008 in the aftermath of a terrorist attack on Mumbai that India viewed as supported by Pakistan. India had the conventional ability to retaliate but chose not to for fear of Pakistani nuclear escalation in response.[38]

In short, India has repeatedly proved willing to forgo much of its conventional military advantage in exchange for a reduced risk of nuclear escalation, suggesting that both it and Pakistan see those risks as closely tied to India's military ability to overrun Pakistan quickly in a conventional war. Counterintuitively, Indian leaders have had to restrain rather than brandish this strength in order to reduce nuclear risks. They clearly recognize a trade-off.[39]

Other powerful combatants have at times demonstrated similar restraint in the conduct of conventional war in order to avoid risks of opponent nuclear escalation. For example, the US was careful to calibrate both its war aims and military operations in the Korean War so that it did not end up "fighting a nuclear World War III against China and the Soviet Union," clearly "a price not worth paying to achieve the goal of long-term South Korean security."[40] The Soviets limited their military interventions in the conflict for the same reason.[41] Again, policy makers based decisions on an implicit understanding of the trade-offs involved in designing a conventional military campaign against an opponent backed by a nuclear-armed superpower.

These patterns call into question the aforementioned stability–instability paradox that guided significant Cold War thinking about nuclear deterrence. The India–Pakistan and Korean War cases, along with the Arab–Israelis Wars of 1967 and 1973, do show that conventional conflict can arise under a nuclear shadow, as the paradox

36 Narang (2014, 270–271). 37 Narang (2014, 275).
38 Narang (2014, 280). See also Ganguly and Kapur (2012, Chapter 4).
39 Powell (2015). 40 Reiter (2009, 64). 41 Carson (2016, 120–124).

would predict, but they also cast doubt on the paradox's notion that there is a clear firebreak between the conventional and nuclear levels, especially if one or both sides do not have the secure second strike forces the paradox assumed, as was the case in the Arab-Israel Wars. Clearly, decision makers in Delhi and Washington perceived nuclear risks to be highly credible, so much so that conventional military operations had to be significantly neutered, ended prematurely, or forgone entirely once nuclear threats were brandished. The level of instability these states were willing to tolerate was very low, posing a real trade-off between conventional military effectiveness and nuclear escalation control.

This trade-off seems likely to persist in the future security environment, particularly if conflicts arise between the US and North Korea, or the US and Russia. North Korea clearly fears conventional attack from the south. The North does not have a formal nuclear doctrine and maintains a highly opaque military establishment, making it hard to analyze escalatory risk on the peninsula in detail. Nevertheless, the country's leaders have made explicit statements suggesting that it would be willing to use nuclear weapons first, much along the lines of NATO and France during the Cold War, and Pakistan (and likely Israel) today.[42] As Keir Lieber and Daryl Press have noted, "War with North Korea probably means nuclear war . . . Ironically, the risk of nuclear war stems not from weakness on the part of the United States and South Korea but from their strength."[43]

Russia, too, is planning for the possible first use of nuclear weapons in the event of an intense conventional conflict with the US or NATO.[44] Although such a conflict remains unlikely, recent tensions over Ukraine and the Baltics, in addition to Russian intervention in Syria, have made war between Russia and the West much more plausible today than at any time since the end of the Cold War. Crucially, recent Russian writings suggest that Russia might use nuclear weapons not only in cases of attack on Russian territory but also in response to conventional attacks on Russian forces engaged in regional conflicts.[45] Put another way, although it remains extremely unlikely that the US or NATO would deliberately conduct military operations directly on Russian territory, Russia has signaled its willingness to use nuclear

[42] Lieber and Press (2013b, 30). [43] Lieber and Press (2013b).
[44] Ven Bruusgaard (2016). [45] Quinlivan and Oliker (2011, 24–25, 27).

weapons short of that threshold, in conflict occurring outside Russia's current borders.

Such conflict could rapidly infringe on what Russia does consider to be its territory. For example, imagine that Russian forces intervened in the Baltics – not a far-fetched possibility given Russian actions in Ukraine and Georgia, and Russia's significant buildup of military capability in adjacent Kaliningrad.[46] Estonia, Latvia, and Lithuania are NATO member states entitled to protection under Article V. Conventionally defending the Baltics might very well require NATO member states, notably the US, to suppress Russian air defense networks and to target command-and-control nodes that are based across the border in Kaliningrad or even in Russia proper.[47] While the US might see these simply as logical targets in a conventional war, the US ability to degrade them so effectively could back Russia into a corner where it believed that brandishing nuclear threats was its best option.

Russia is reportedly forward deploying nuclear-capable missiles in Kaliningrad, possibly as a way of signaling that it sees threats to Kaliningrad in the most serious possible terms. The key thing to note is that this sort of risk could arise not because of a foolish US attempt to attack Russian territory directly but from the plausible course of effective military operations in defense of a treaty ally. As a recent RAND study notes, "Russia has made it clear that, given its conventional inferiority to plausible adversaries, including the United States and NATO, it might be forced to use nuclear weapons in response to a conventional attack."[48] Certainly Russian leaders have never disavowed such use, and Russian statements and writings suggest that in cases where the country faced what it considered existential dangers from a conventional onslaught, it might respond with nuclear weapons.[49]

Russian military exercises reflect these efforts to maintain a nuclear first-strike capability, especially given the increasingly apparent constraints on Russia's second strike capabilities.[50] According to the RAND study, Russian political leaders have "endorsed military

[46] Radio Free Europe/Radio Liberty (2015).
[47] Fruhling and Lasconjarias (2016); Colby and Solomon (2015).
[48] Quinlivan and Oliker (2011, ix).
[49] Norberg (2015); Quinlivan and Oliker (2011, 18–19).
[50] Lieber and Press (2006, 2013c). The RAND study notes that Russia continues to pursue the possibility of a dead hand system that could retaliate even in the event of a decapitating first strike. Quinlivan and Oliker (2011, 33).

doctrines that allowed for first use of nuclear weapons and conspicuously endorsed military exercises that included launch of strategic systems through 2009."[51] Notably, exercises in March 2015 simulated "a full-scale confrontation with NATO through the forward deployment of nuclear-armed submarines, theater ballistic missiles, and strategic bomber aircraft. Strategic weapons systems, including assets that are part of Russia's nuclear capabilities, have also been deployed to locations near NATO's borders."[52]

Russian first use of nuclear weapons during a conventional war could take several possible forms.[53] Some Russian writings suggest that a strategy of escalation might begin with demonstration strikes in sparsely populated areas, as part of a coercion effort.[54] Alternatively, or in addition, Russia might use theater nuclear weapons as a means of staunching a conventional attack and pressuring its adversary to back down. In fact, some Russian writers view nuclear weapons as a means of escalating to force deescalation.[55]

Although the overall probability of war remains low, and the probability of nuclear escalation within that war lower still, it is hard to dismiss outright. Counterintuitively, the military effectiveness of NATO's conventional military operations could raise the risk of Russian first use of nuclear weapons.

Nuclear Escalation as a Response to Conventional Counterforce

Even a nuclear-armed state that did not fear impending conventional defeat might feel pressure to resort to nuclear weapons in the event that an opponent's conventional military operations threatened the state's nuclear capabilities. This notion of "conventional counterforce" – of an opponent's conventional military operations having the ability to threaten the state's nuclear arsenal – represents a second pathway by which conventional military effectiveness against a nuclear-armed opponent could lead to opponent nuclear escalation.

[51] Quinlivan and Oliker (2011, x). [52] STRATFOR (2015).

[53] It is important note that these arguments are probabilistic and that Russian doctrine may not be as coherent as is often assumed in the West. Adamsky (2014).

[54] Quinlivan and Oliker (2011, 30). [55] Quinlivan and Oliker (2011, 28).

Nuclear weapons are often viewed as stabilizing because they are not very useful for conquering others' territory, but they are very helpful in deterring attacks on one's own.[56] Precisely because of the importance states attach to their nuclear arsenals, however, they tend to treat threats to those arsenals with the highest possible concern. Conventional military operations that implicate the security of a state's nuclear arsenal are likely to be viewed as highly threatening and offensive, even when such operations are designed to be defensive or to be offensive only from the perspective of conventional rather than nuclear war.[57] Put another way, the threat *of* nuclear weapons may inhibit escalation from peace to war, but threats *to* nuclear weapons may provide reasons for intrawar escalation once war has broken out.[58]

At least three types of nuclear or nuclear-relevant targets could come under attack in conventional wars. First, in situations where nuclear forces were physically co-located with conventional ones, attacks on conventional targets could also threaten the survivability of a state's nuclear arsenal. For example, if a state kept nuclear and conventional missiles at the same bases or intermixed them in mobile launch brigades, an opponent's effort to eliminate the conventional missile threat as part of a normal, effective conventional campaign could also start to erode the state's nuclear capabilities.

Second, some weapons systems and platforms are relevant to both nuclear and conventional missions. For example, dual-capable bombers would be an obvious target of choice in a conventional air campaign, but attacking those bombers even when they were only conventionally armed could also indirectly erode the state's nuclear deterrent capacity (though to what extent would depend on other factors). Likewise, attack submarines are critical to the conventional defense of sea lanes, but they are also crucial for protecting ballistic missile submarines, which could become more vulnerable in the event that attack submarines were sunk.

Third, command-and-control (C2) networks for nuclear and conventional systems have the potential to overlap substantially. Disrupting an adversary's ability to gather and disseminate information and to issue military orders is a central goal of effective conventional military campaigns, yet in situations where nuclear and conventional C2

[56] Glaser (1990); Jervis (1989); Sagan and Waltz (1997).
[57] Posen (1991, 13–15). [58] Talmadge (2017b).

overlap, military commanders and political leaders might be at risk of losing the ability to use not only their conventional but also their nuclear forces. For example, air defense radars can be used not only for a conventional air war but also as part of a state's nuclear early warning system. A state that relied on a launch-on-warning posture might therefore view the destruction of these radars as far more than an attack on air defenses. Losing these radars could render its nuclear forces dramatically more vulnerable to a first strike. As a result, an opponent might face a trade-off between maximizing conventional military effectiveness, which would call for destroying the radars, and minimizing nuclear escalatory risk, which might call for leaving the radars intact even though they could make the opponent's conventional campaign much more costly.

Conventional military attacks on these sorts of nuclear-relevant targets could plausibly occur through sheer inadvertence; one side could simply fail to appreciate the nuclear implications of its conventional operations. For example, Barry Posen argued that in the late 1970s and 1980s political leaders in the US and NATO failed to recognize the nuclear counterforce pressures that their conventional military operations would have placed on the Soviets in the event of war in Central Europe.[59] The US maritime strategy at the time called for a highly offensive approach to sea control in order to secure the sea lines of communication that would have ferried reinforcements to Europe in the event of a conventional war. This offensive approach would have required early and aggressive efforts by the US Navy to track and sink Soviet submarines in the Norwegian and Barents seas, as well as to destroy Soviet shore-based naval infrastructure. There was a high probability that such efforts – if effective at the tactical and operational levels of war – would have resulted in the sinking of Soviet ballistic missile submarines (SSBNs). Even if Soviet SSBNs had not been sunk, the Soviets surely would have feared such a prospect once the Soviet attack submarine (SSN) force had suffered significant attrition and Soviet shore-based naval infrastructure had been destroyed.[60]

Similarly, many Soviet early warning radars and air defenses were forward deployed in areas that NATO surely would have attacked early in a conflict merely to establish control of the air for the conventional campaign. As Posen notes, the Soviets would have been on

[59] Posen (1991). [60] Posen (1991, Chapters 1 and 4); Mearsheimer (1986).

guard for signs of an impending nuclear attack. Even if the US did not intend to launch one, "ambiguous evidence in wartime suggesting that the United States could have beaten [the Soviets] to the punch could have been the spark that set off their nuclear offensive."[61]

Under such circumstances, the Soviets could have turned to nuclear weapons for a few purposes. First, if the Soviets were truly convinced that these conventional counterforce attacks meant that the US was gearing up for a nuclear first strike, the Soviets could have launched a full-on nuclear first strike of their own for purposes of damage limitation – that is, for the purpose of trying to destroy the US nuclear force before it could be used against the Soviet force. Such a strike would have been insane during peacetime, of course, but under wartime conditions where Soviet leaders believed nuclear war was inevitable, they might have seen such escalation as the least bad option.

Second, and probably more realistically, Soviet leaders could have used their nuclear weapons in more limited ways, especially if they did not really believe that the US was truly committed to and preparing for a first strike. For example, the Soviets could have used nuclear weapons to immediately halt the components of the US conventional campaign that posed the most significant threats to the Soviet nuclear forces, for example, through tactical strikes against carrier task forces or naval bases supporting operations that were inadvertently degrading the Soviet SSBN force. The Soviets also could have conducted demonstration strikes against isolated military targets or even in unpopulated areas to signal to the US that the conventional campaign was crossing a Soviet red line and was started to risk larger-scale nuclear escalation.

The key point is that nuclear escalation would not necessarily have arisen because the Soviets were losing the conventional war, but because of the perceived effects of the conventional campaign on the Soviet nuclear deterrent. Moreover, these dangerous nuclear consequences would have been the direct result of the effective conduct of US conventional military operations.[62] The better the US had been at executing its plans for conventional victory at sea, on land, and in the air in the late Cold War, the more likely it was that the Soviets would have

[61] Posen (1991, Chapters 2 and 3, and 66). See also Ball (1981); Blair (1985).

[62] In fact, the US may have knowingly pursued this escalatory conventional strategy in a belief that it conferred additional general deterrence benefits or bargaining advantages. Mearsheimer (1986); Jervis (1989, Chapter 5); Powell (2015, 593–597); Talmadge (2017a).

experienced conventional counterforce fears that could have prompted them to use nuclear weapons first.

Similar dynamics could arise today with respect to potential nuclear escalation in a US–China conflict.[63] Certainly, the overall risk of war between the two countries is low. Nevertheless, sources of tension remain.[64] In the event that a conventional war were to arise, some analysts have raised concerns about the possibility that US conventional military operations might create inadvertent nuclear pressures on China – particularly given that prevailing US concepts of operation likely would involve extensive attacks on the Chinese mainland, some of which could involve nuclear-relevant targets.[65] Worryingly, Chinese writings have emphasized that conventional attacks on China's strategic nuclear forces might be treated as the equivalent of nuclear attacks, vitiating China's no-first-use pledge and justifying nuclear use by China.[66] For example, Avery Goldstein argues that the use of conventional force is inherently unpredictable and could lead to a variety of miscalculations between the US and China that might eventually result in "unanticipated nuclear catastrophe." In his view, a particular danger stems from the possibility that the US might inadvertently sink a Chinese SSBN in the course of conventional military operations, thereby "inviting Chinese nuclear retaliation."[67]

Thomas Christensen similarly emphasizes that the conventional-nuclear firebreak might not be sufficiently robust in the event of a US–China war, because "China is simultaneously developing conventional and nuclear coercive capabilities that overlap significantly." He points in particular to the dual nuclear and conventional relevance of Chinese submarines, missiles, and command-and-control systems, arguing that US conventional attacks on such targets "could be mistaken in Beijing as part of an attempted US first strike on China's nuclear deterrent," resulting in pressure to escalate.[68]

Even a recent RAND study that is less alarmist about nuclear dangers in general still concludes that China's "conventionally armed DF-21s [a medium-range ballistic missile] may be difficult or impossible to distinguish from nuclear-armed DF-21s, and the hunt for

[63] Talmadge (2017b). [64] Liff and Ikenberry (2014); Glaser (2015).
[65] Rovner (2012); Manzo (2015).
[66] Fravel and Medeiros (2010, 80); Christensen (2012, 454, 475–477).
[67] Goldstein (2013a). See also Goldstein (2013b, 70–72).
[68] Christensen (2012, 453–454, 468, esp. note 54, 472, 476).

conventionally armed missiles could result in the attrition of China's nuclear-capable missile force." The authors warn that "this could ultimately create a 'use-them-or-lose-them' dilemma for Chinese strategic planners, particularly if other parts of China's strategic system (such as SSBNs) were under attack."[69] In fact, this concern has led some analysts to propose more defensive US concepts of operation for any potential war in the Western Pacific, simply because they view the risks inherent in conventional attacks on the Chinese mainland or even within the first island chain as so significant.[70]

Clearly, no matter how threatened it is, China lacks the ability to use nuclear weapons in an attempt at a damage limitation strike against the US as the Soviets did. However, China still could see nuclear weapons as useful for instantaneously halting components of a US conventional campaign that were posing the greatest threat to China's nuclear arsenal, or for coercively signaling to the US that it needed to reconsider an impending attempt at conventional counterforce.

Not all observers are concerned, however. For example, Elbridge Colby, a former Pentagon official, recognizes that a conventional US campaign would attack targets on the Chinese mainland but believes there is little reason to believe that this approach would be "likely to spur a Chinese nuclear response."[71] He does not dismiss escalatory risk entirely but argues that US policy makers are sensitized to such dangers and are able to manage them properly – a view that US policy makers themselves have expressed repeatedly in interviews.[72]

Interestingly, Chinese strategists have been similarly sanguine, as reported by Fiona Cunningham and Taylor Fravel in a recent study drawing on extensive Chinese-language sources and interviews.[73] They note that "overall, Chinese strategists do not see a high risk of unintentional escalation, believing instead that the United States and China would tightly control any crisis or conflict to prevent escalation." While not accepting these Chinese views uncritically, Cunningham and Fravel note that "open-source materials indicate that the majority of China's

[69] Heginbotham et al. (2015, 353–355). See also Riqiang (2014); Twomey (2011, 64).

[70] Hammes (2012); Kline and Hughes (2012). [71] Colby (2014b).

[72] Discussions and interviews conducted with nearly three dozen current and former midlevel to senior civilians and military officers with experience serving in the Department of Defense, Department of State, and Intelligence Community, fall 2014 to summer 2016. Participants requested anonymity.

[73] Cunningham and Fravel (2015).

nuclear missiles are not co-located with conventional ones."[74] Cunningham and Fravel also share some of the doubt expressed by their Chinese sources and interlocutors about whether US attacks on China's command-and-control system for conventional assets would impinge on China's command-and-control system for nuclear assets. As they explain, "Although conventional and nuclear launch brigades may share some support regiments at the base level, including warhead transportation and communications regiments, and may use some of the same physical infrastructure for command and control, they use different command and control chains." In addition, they argue that Chinese command-and-control systems are more capable and redundant than often realized in Western analyses.[75]

If they are right, then the trade-off the US faces in a potential conflict with China between conventional military effectiveness and nuclear escalation risks may be less intense than comparable trade-offs that would have been involved in fighting the Soviets during the Cold War. China may simply have fewer nuclear-relevant assets that could be ensnared in the course of conventional military operations, even fairly expansive ones on the Chinese mainland. Most importantly, if China believes that its land-based, increasingly mobile intercontinental ballistic missile (ICBM) force is likely to remain secure regardless of the course of conventional conflict with the US, it would have little reason to escalate. For example, even if the US inadvertently destroyed nuclear-tipped versions of the DF-21, much of China's retaliatory capability would remain intact and far inland, away from likely fighting in coastal regions.

A critical question is whether Chinese leaders see it this way, and whether they would view such attacks, in the context of an ongoing conventional war and in combination with other possible attacks on potentially nuclear-relevant assets, as the prelude to a larger counterforce strike. Chinese worries about US missile defense in the region imply worries about this larger possibility – that the US is seeking a damage limitation capacity because it wants to retain the option for a first strike against China.[76]

Much also depends on how survivable Chinese leaders believe their arsenal is. Open-source assessments vary significantly. Some analysts

[74] Cunningham and Fravel (2015, 42). [75] Cunningham and Fravel (2015, 44).

[76] On Chinese views of US missile defenses in the region, see Riqiang (2011, 2013); Li (n.d.); Twomey (2013, 295). On damage limitation, see Glaser and Fetter (2016).

believe that China's mobile missiles are highly survivable, meaning the US would have a hard time hunting them even if it wanted to, much less threatening them inadvertently.[77] Hence, there would be little reason to expect China to experience escalatory pressures if it is behaving rationally. By contrast, other analysts suggest that China's arsenal is quite vulnerable, because of both its small size and the US's increasingly sophisticated ability to hunt mobile targets, especially large ICBMs whose movements would be somewhat predictable and whose preparations for use would generate signals visible to US intelligence.[78] If this is true, then China would have reason to feel fairly nervous in the event that it detected any warning signs, even purely conventional ones, that the US might be preparing for a larger attempt at counterforce.

In a perfect world the US could easily signal to China that it did not intend to conduct such a campaign, easing Chinese fears, but the empirical record does not induce optimism on this score. In general US–China crisis/wartime communications channels are very limited – much less developed than they were with the Soviet Union, for example.[79] Simulations also have shown that both the US and China tend to envision themselves acting defensively and with limited aims, and that they see escalation as something that would result from the other side's choices, not their own, so the lack of communications channels to clarify matters could be particularly pernicious.[80]

Furthermore, even in past wars where US political leaders have had limited aims, US military campaigns have not always successfully communicated those clearly to the adversary – even when that adversary had weapons of mass destruction (WMD). For example, in the 1991 Gulf War, the Bush administration explicitly did not seek to overthrow the regime of Iraqi dictator Saddam Hussein (though it was not opposed to creating the conditions for his removal by domestic opponents). Nevertheless, as an authoritative history of the war notes, "For the air-war planners, Saddam Hussein had been a target of opportunity and the destruction of his regime a goal." Up until the final hours of the war, in fact, their massive air campaign targeted a variety of regime sites and individuals, irrespective of the fact that US political leaders had not identified regime change as a goal.[81]

[77] Li (2007).
[78] Lieber and Press (2006); Riqiang (2013, esp. 587, 591); Long and Green (2014).
[79] Goldstein (2013b). [80] Roberts (2007, 230).
[81] Gordon and Trainor (1995, 137, 408, 414, 418, 423, 442).

This lack of US military restraint is even more striking given Iraq's stocks of chemical and biological weapons at the time, and the country's history of extensively using the former in the recent war with Iran.[82] In addition to the regime targets already mentioned, the air campaign vigorously attacked suspected Iraqi WMD sites deep in Iraqi territory, including sites related to the nuclear program, again without apparent fear that this might make Iraq more rather than less likely to use such weapons, about which Western intelligence was murky at the time.[83]

Surprisingly, despite the known presence of WMD, there is virtually no evidence in interviews with or memoirs of senior US officials that these capabilities figured into either US civilian or military decision making about how to communicate war aims, how to conduct the war, or when to declare the war over.[84] Several explanations for this apparent lack of US concern are possible, including the belief that the US nuclear deterrent would prevent Saddam from using WMD regardless of the extent of US conventional operations, as well as possible differing assessments of how capable the Iraqi WMD arsenal actually was. In the aftermath of the war, both Iraqi and US officials claimed that Saddam had refrained from using chemical and biological weapons because of ambiguous US threats of "the strongest possible response" to such attacks.[85] Secretary of State James Baker had delivered this threat via a letter from President Bush to Tariq Aziz in a meeting in January 1991.

However, in retrospect there are several reasons to doubt that Saddam's restraint in not using BW and CW stemmed from effective US deterrent threats, casting some doubt on whether comparable threats would serve to restrain future nuclear-armed adversaries considering nuclear escalation.[86] First, despite Bush's much-lauded letter, US threats against Saddam were actually highly ambiguous along several dimensions, making it hard to know exactly how Iraq would have interpreted them.[87] Second, the letter from Bush promised "the strongest possible response" not only in the event of an Iraqi WMD attack, but also if Iraqi engaged in terrorist attacks in the region or burned Kuwait's oil wells – both actions Iraq did eventually take. These

82 Cordesman and Wagner (1991); Talmadge (2013, 2015, Chapters 4 and 5).
83 Gordon and Trainor (1995, 255–256, 327, 460).
84 Gordon and Trainor (1995, 404–405, 415–416).
85 Woods et al. (2011, 221); Sagan (2000, 91).
86 Sagan (2000, 91).
87 Sagan (2000, 91–92).

choices call into question whether Iraqi leaders really found the threats in the letter credible.[88] Third, the trove of Iraqi government documents and recordings obtained after the 2003 war provides no evidence that the US threats played a role in the Iraqi decision not to use WMD. They do show, though, that the Iraqi leadership essentially saw American aims as unlimited (a reasonable inference from the air campaign described above), and that Iraq did not see its WMD capability as having any ability to temper US intentions to pursue the war all the way to Baghdad.[89] Hence it is hard to make the case either that the US successfully conveyed its limited war aims to its opponent, or that the opponent refrained from using WMD because of the limited US war aims – not an encouraging finding.

Instead, Saddam most likely refrained from using chemical and biological weapons not because of clear US communication of its limited war aims, or because of US threats of nuclear retaliation, but for practical reasons. Iraq's best means of delivering chemical munitions – aircraft and artillery – had been heavily damaged by the air war by the time coalition ground forces were anywhere where Iraq might have attacked them with chemical weapons. In addition, weather conditions in the theater would not have been favorable to chemical weapons use, something the Iraqis would have known from their past experience using chemical weapons in the Iran-Iraq War. Finally, Iraq probably believed based on media reports that coalition soldiers had protective equipment that would have nullified the effects of any chemical attack, and that coalition operations (unlike those of the Iranians in the previous decade) were too mobile to have made chemical attack very effective.[90]

For all of these reasons, the case does not lend itself to optimism about the future US ability to signal limited aims even against adversaries armed with WMD and adversaries' ability to interpret those limits clearly and to restrain their behavior accordingly. The firebreak between the conventional and nuclear realms may not be as strong as the stability–instability paradox would imply, especially if the paradox's assumption of mutual vulnerability is not met, as may be the case with new or small nuclear powers. As such, the US and other countries may face wrenching trade-offs as they seek to reduce nuclear escalatory risk with pursuing conventional military effectiveness.

[88] Woods et al. (2011, 221–222).
[89] Woods et al. (2011, 236–236); Woods (2008, esp. Chapter 7).
[90] Cigar (1992, 151).

Conclusions and Implications

If military effectiveness is about the ability to achieve a given set of political goals at acceptable costs, then escalation control is a fundamental component of effectiveness. Nowhere is this concern more consequential than in conventional wars between nuclear-armed states. Nuclear-armed states do at times engage in conventional conflict, and the nuclear–conventional firebreak is not as robust as often assumed. Future nuclear escalation is still unlikely, but it is not impossible, and the risk of opponent nuclear escalation actually can rise as a military's tactical and operational effectiveness increases.

Clearly, the threat of conventional defeat on the battlefield has raised risks of deliberate nuclear escalation in past wars involving India and Pakistan, and Israel and the Arab states. It is plausible that such risks also could arise in future confrontations between Russia and NATO, or between North Korea and US-Republic of Korea coalition forces.

Nuclear escalation also could arise more inadvertently, in situations where one state's conventional military campaign started to infringe on its opponent's nuclear deterrent. This risk clearly existed in the Cold War and plausibly could arise again today in conventional conflict between the US and China. Although the precise nature of the risks in a China scenario are different than they were with the Soviet Union, it is hard to dismiss them outright. The historical record indicates that in at least some cases states have had difficulty convincing their adversaries that their goals did not involve winner-take-all objectives. Nuclear use did not occur in those cases, but the specific reasons behind nonuse in those cases will not necessarily be present in future crises or wars.

What should policy makers do when facing this potential tradeoff between conventional military effectiveness and nuclear escalatory danger? Policy makers could circumscribe the components of a conventional military campaign that might pose the most direct threat to an opponent's territorial integrity or nuclear arsenal. This approach could have serious costs, not just militarily but politically, but might be worthwhile, particularly if there are alternative conventional concepts of operations that can achieve the state's objectives with lower nuclear risks.[91]

[91] On India's attempt to solve this problem, see Powell (2015, 613).

Policy makers also might consider attempting to ease the trade-off by shaping the adversary perceptions that could lead to a decision to escalate. Regarding the first pathway, for example, involving states that threaten deliberate nuclear escalation in the event of conventional losses, policy makers could signal a willingness to respond in kind, perhaps against the very targets that the escalating state might be escalating to preserve (although this move clearly would have other costs). Regarding the second pathway, involving states that might escalate because of fears of conventional counterforce, policy makers might attempt to emphasize, especially through prewar signals, that their aims are limited and that conventional infringement on nuclear targets does not signal a broader set of counterforce objectives. Making such reassurances credible, especially once a large-scale conventional war is underway, could prove very difficult but is worth attempting given the stakes involved.

References

Adamsky, Dmitry. 2014. Nuclear Incoherence: Deterrence Theory and Non-Strategic Nuclear Weapons in Russia. *Journal of Strategic Studies* 37(1): 91–134.

Ball, Desmond. 1981. *Can Nuclear War Be Controlled*?Adelphi Paper 169. London: International Institute for Strategic Studies.

Blair, Bruce. 1985. *Strategic Command and Control: Redefining the Nuclear Threat*. Washington, DC: Brookings Institution Press.

1993. *The Logic of Accidental Nuclear War*. Washington, DC: Brookings Institution Press.

Carson, Austin. 2016. Facing Off and Saving Face: Covert Intervention and Escalation Management in the Korean War. *International Organization* 70(Winter): 103–131.

Carter, Ashton, John D. Steinbruber, and Charles A. Zraket, eds. 1987. *Managing Nuclear Operations*. Washington, DC: Brookings Institution Press.

Castillo, Jasen. 2014. Deliberate Escalation: Nuclear Strategies to Deter or to Stop Conventional Attacks. Unpublished manuscript, Texas A&M University.

Christensen, Thomas. 2012. The Meaning of the Nuclear Evolution: China's Strategic Modernization and US-China Security Relations. *Journal of Strategic Studies* 35(August): 447–487.

Cigar, Norman. 1992. Chemical Weapons and the Gulf War: The Dog That Did Not Bark. *Studies in Conflict and Terrorism* 15(2): 145–155.

Cohen, Avner. 1999. *Israel and the Bomb*. New York: Columbia University Press.

Colby, Elbridge. 2014a. The United States and Discriminate Nuclear Options in the Cold War. In Jeffrey A. Larsen and Kerry M. Kartchner, eds., *On Limited Nuclear War in the 21st Century*, 49–79. Stanford, CA: Stanford University Press.

2014b. Welcome to China and America's Nuclear Nightmare. *The National Interest*, December 19.

Colby, Elbridge, Avner Cohen, William McCants, Bradley Morris, and William Rosenau. 2013. *The Israeli Nuclear Alert of 1973: Deterrence and Signaling in Crisis*. Washington, DC: Center for Naval Analyses.

Colby, Elbridge, and Jonathan Solomon. 2015. Facing Russia: Conventional Defense and Deterrence in Europe. *Survival* 57(6).

Cordesman, Anthony, and Abraham Wagner. 1991. *The Lessons of Modern War: Vol. 2. The Iran-Iraq War*. Boulder, CO: Westview.

Cunningham, Fiona, and Taylor Fravel. 2015. Assuring Assured Retaliation: China's Nuclear Strategy and US-China Strategic Stability. *International Security* 40(Fall): 7–50.

Forrest Morgan E., Karl P. Mueller, Evan S. Medeiros, Kevin L. Pollpeter, and Roger Cliff. 2008. *Dangerous Thresholds: Managing Escalation in the 21st Century*. Santa Monica, CA: RAND Corporation.

Fravel, Taylor, and Evan Medeiros. 2010. China's Search for Assured Retaliation: The Evolution of Chinese Nuclear Strategy and Force Structure. *International Security* 35(Fall): 48–87.

Fruhling, Stephan, and Guillaume Lasconjarias. 2016. NATO, A2/AD, and the Kaliningrad Challenge. *Survival* 58(2): 95–116.

Ganguly, Sumit, and S. Paul Kapur. 2012. *Indian, Pakistan, and the Bomb: Debating Nuclear Stability in South Asia*. New York: Columbia University Press.

Gavin, Francis J. 2012. *Nuclear Statecraft: History and Strategy in America's Atomic Age*. Ithaca, NY: Cornell University Press.

Glaser, Charles L. 1990. *Analyzing Strategic Nuclear Policy*. Princeton, NJ: Princeton University Press.

2015. A US-China Grand Bargain? The Hard Choice between Military Competition and Accommodation. *International Security* 39(Spring): 49–90.

Glaser, Charles L. and Steve Fetter. 2016. Should the United States Reject MAD? Damage Limitation and U.S. Nuclear Strategy toward China. *International Security* 41(Summer): 49–98.

Goldstein, Avery. 2013a. China's Real and Present Danger: Now Is the Time for Washington to Worry. *Foreign Affairs* (September/October): 136–144.

2013b. First Things First: The Pressing Danger of Crisis Instability in US-China Relations. *International Security* 37(Spring): 49–89.

Gordon, Michael R., and Bernard E. Trainor. 1995. *The Generals' War: The Inside Story of the Conflict in the Gulf*. New York: Little, Brown.

Green, Brendan Rittenhouse, and Austin G. Long. 2015. Counterforce, Private Information, and Cross-Domain Deterrence: Evidence on Soviet Perceptions and Policy during the Late Cold War. Unpublished manuscript, Williams College.

Hammes, T. X. 2012. *Offshore Control: A Proposed Strategy for an Unlikely Conflict*. Strategic Forum No. 278. Washington, DC: Institute for National Strategic Studies, National Defense University.

Heginbotham, Eric, et al. 2015. *The US-China Military Scorecard: Forces, Geography, and the Evolving Balance of Power, 1996–2017*. Arlington, VA: RAND Corporation.

Jervis, Robert. 1979–1980. Why Nuclear Superiority Doesn't Matter. *Political Science Quarterly* 94(Winter): 617–633.

1984. *The Illogic of American Nuclear Strategy*. Ithaca, NY: Cornell University Press.

1989. *The Meaning of the Nuclear Revolution: Statecraft and the Prospect of Armageddon*. Ithaca, NY: Cornell University Press.

Kissinger, Henry. 1995. *Diplomacy*. New York: Simon and Schuster.

Kline, Jeffrey, and Wayne Hughes. 2012. Between Peace and the Air-Sea Battle: A War at Sea Strategy. *Naval War College Review* 65(4): 34–41.

Krepinevich, Andrew. 1988. *The Army and Vietnam*. Baltimore: Johns Hopkins University Press.

Kroenig, Matthew. 2013. Nuclear Superiority and the Balance of Resolve: Explaining Nuclear Crisis Outcomes. *International Organization* 67(Winter): 141–171.

Li, Bin. 2007. Tracking Chinese Strategic Mobile Missiles. *Science and Global Security* 15: 1–30.

N.d. The Impact of the US NMD on the Chinese Nuclear Modernization. Working paper, Institute of Science and Public Affairs, China Youth College for Political Science.

Lieber, Keir, and Daryl Press. 2006. The End of MAD? The Nuclear Dimension of US Primacy. *International Security* 30(Spring): 7–44.

2013a. Coercive Nuclear Campaigns in the 21st Century: Understanding Adversary Incentives and Options for Nuclear Escalation. PASCC Report 2013-001, Naval Postgraduate School.

2013b. The Next Korean War. *Foreign Affairs Snapshot*, April 1.

2013c. The New Era of Nuclear Weapons, Deterrence, and Conflict. *Strategic Studies Quarterly* 7(Spring): 3–14.

Liff, Adam, and G. John Ikenberry. 2014. Racing toward Tragedy? China's Rise, Military Competition in the Asia Pacific, and the Security Dilemma. *International Security* 39(Fall): 52–91.

Long, Austin, and Brendan Rittenhouse Green. 2014. Stalking the Secure Second Strike: Intelligence, Counterforce, and Nuclear Strategy. *Journal of Strategic Studies* 38(1–2): 38–73.

Manzo, Vincent. 2015. After the First Shots: Managing Escalation in Northeast Asia. *Joint Forces Quarterly* 77(2): 91–100.

Mearsheimer, John. 1986. A Strategic Misstep: The Maritime Strategy and Deterrence in Europe. *International Security* 11(Fall): 3–57.

Narang, Vipin. 2014. *Nuclear Strategy in the Modern Era: Regional Powers and International Conflict*. Princeton, NJ: Princeton University Press.

New York Times. 2015. Opening New Iraq Front, US Strikes ISIS in Tikrit. March 25.

Norberg, Johan. 2015. *Training to Fight: Russia's Major Military Exercises 2011–2014*. Stockholm: FOI.

Posen, Barry. 1991. *Inadvertent Escalation: Conventional War and Nuclear Risks*. Ithaca, NY: Cornell University Press.

Powell, Robert. 1985. Theoretical Foundations of Strategic Nuclear Deterrence. *Political Science Quarterly* 100(Spring): 75–96.

1987. Crisis Bargaining, Escalation, and MAD. *American Political Science Review* 81(3): 717–736.

2015. Nuclear Brinkmanship, Limited War, and Military Power. *International Organization* 69(Summer): 589–626.

Quinlivan, James, and Olga Oliker. 2011. *Nuclear Deterrence in Europe: Russian Approaches to a New Environment and Implications for the United States*. Arlington, VA: RAND Corporation.

Radio Free Europe/Radio Liberty. 2015. Kaliningrad, Moscow's Military Trump Card. October 29. www.rferl.org/content/kaliningrad-russia-nato-west-strategic/27079655.html.

Reiter, Dan. 1995. Exploding the Powder Keg Myth: Preemptive Wars Almost Never Happen. *International Security* 20(Fall): 5–34.

2009. *How Wars End*. Princeton, NJ: Princeton University Press.

2014. Security Commitments and Nuclear Proliferation. *Foreign Policy Analysis* 10(1): 61–80.

Riqiang, Wu. 2011. Survivability of China's Sea-Based Nuclear Forces. *Science and Global Security* 19(2011): 91–120.

2013. Certainty of Uncertainty: Nuclear Strategy with Chinese Characteristics. *Journal of Strategic Studies* 36(May): 579–614.

2014. Sino-US Inadvertent Escalation. Unpublished manuscript.

Roberts, Brad. 2007. The Nuclear Dimension: How Likely? How Stable? In Michael Swaine, Andrew Yang, and Evan Medeiros, eds., *Assessing the*

Threat: The Chinese Military and Taiwan's Security. Washington, DC: Carnegie Endowment.

Rovner, Joshua. 2012. AirSea Battle and Escalation Risks. Policy Brief 12, University of California Institute on Global Conflict and Cooperation.

Sagan, Scott. 1989. *Moving Targets: Nuclear Strategy and National Security*. Princeton, NJ: Princeton University Press.

1995. *The Limits of Safety: Organizations, Accidents, and Nuclear Weapons*. Princeton, NJ: Princeton University Press.

2000. The Commitment Trap: Why the United States Should Not Use Nuclear Threats to Deter Biological and Chemical Weapons Attacks. *International Security* 24(Spring): 85–115.

Sagan, Scott, and Kenneth Waltz. 1997. *The Spread of Nuclear Weapons: A Debate*. New York: W. W. Norton.

Schelling, Thomas. 1980. *The Strategy of Conflict*. Cambridge, MA: Harvard University Press.

Sechser, Todd, and Matthew Fuhrmann. 2013. Debating the Benefits of Nuclear Superiority for Crisis Bargaining, Part II. *Duck of Minerva* (blog), March 25.

Shirkey, Zachary C. 2009. *Is This a Private Fight or Can Anybody Join? The Spread of Interstate War*. Burlington, VT: Ashgate.

Smoke, Richard. 1977. *War: Controlling Escalation*. Cambridge, MA: Harvard University Press.

Snyder, Glenn. 1961. *Deterrence and Defense: Toward a Theory of National Security*. Princeton, NJ: Princeton University Press.

1965. The Balance of Power and the Balance of Terror. In Paul Seabury, ed., *The Balance of Power*, 184–201. San Francisco: Chandler.

Snyder, Jack. 1984. *The Ideology of the Offensive: Military Decision Making and the Disasters of 1914*. Ithaca, NY: Cornell University Press.

STRATFOR. 2015. Russia Targets NATO with Military Exercises. Brief. March 19.

Summers, Harry. 2009. *On War: A Critical Analysis of the Vietnam War*. New York: Presidio Press.

Talmadge, Caitlin. 2013. The Puzzle of Personalist Performance: Iraqi Battlefield Effectiveness in the Iran-Iraq War. *Security Studies* 22(2): 180–221.

2015. *The Dictator's Army: Battlefield Effectiveness in Authoritarian Regimes*. Ithaca, NY: Cornell University Press.

2017a. Risky Business: Nuclear Dangers in Conventional Wars. Unpublished manuscript.

2017b. Would China Go Nuclear? Assessing the Risk of Chinese Nuclear Escalation in a Conventional War with the United States. *International Security* 40(Spring): 50–92.

Twomey, Christopher. 2011. Asia's Complex Strategic Environment: Nuclear Multipolarity and Other Dangers. *Asia Policy* 11(January): 51–78.

2013. Nuclear Stability at Low Numbers: The Perspective from Beijing. *The Nonproliferation Review* 20(2): 289–303.

Ven Bruusgaard, Kristin. 2016. Russian Strategic Deterrence. *Survival* 58(4): 7–26.

Woods, Kevin. 2008. *The Mother of All Battles: Saddam Hussein's Strategic Plan for the Persian Gulf War*. Annapolis, MD: Naval Institute Press.

Woods, Kevin, David D. Palkki, and Mark E. Stout. 2011. *The Saddam Tapes: The Inner Workings of a Tyrant's Regime, 1978–2001*. New York: Cambridge University Press.

8 *Making Trade-offs without Assessing Probabilities*

The Costs and Benefits of Vague Information in National Security Decision Making

JEFFREY A. FRIEDMAN

The central theme of this volume is that the pursuit of military effectiveness requires making trade-offs. It is impossible to evaluate those trade-offs without assessing probabilities. Logically speaking, a decision maker cannot justify any course of action without believing that its chances of success are high enough to make expected benefits exceed expected costs. Carl von Clausewitz thus captured the central role that uncertainty plays in war by writing that "no other human activity is so continuously or universally bound up with chance... From the very start there is an interplay of possibilities, probabilities, good luck and bad that weaves its way throughout the length and breadth of the tapestry."[1]

Yet assessing uncertainty in war is not like forecasting sporting events or weather patterns, where analysts can draw on large volumes of well-behaved data in order to generate rigorous probability estimates. The most important questions about armed conflicts and world politics involve deep uncertainty, and any attempts to resolve this uncertainty are inherently subjective.[2] Many national security agencies thus instruct intelligence analysts and military planners to leave their probabilistic assumptions deliberately vague, even when analyzing basic issues like the risks that military operations pose and the chances that those operations will succeed.

Vague treatments of probability can pose major challenges to making sound decisions. For example, when President John F. Kennedy was considering the Bay of Pigs invasion in 1961, he asked the Joint Chiefs

[1] Clausewitz (1976, 85–86).

[2] Beyerchen (1992–1993); Jervis (1997); Betts (2000); Watts (2004); Connable (2012).

of Staff to evaluate the plan's feasibility. In internal discussions, the Joint Chiefs concluded that the invasion's odds of succeeding were only three-in-ten. But the Joint Chiefs felt awkward conveying this subjective judgment explicitly. Thus in their final report, they stated only that "this plan has a fair chance of success." The Joint Chiefs intended to use the word "fair" in much the same way that a letter grade of C reflects "fair" performance on a test: something that could use a lot of improvement. But President Kennedy misinterpreted that statement to indicate optimism, as if the word "fair" really meant "fine." The report's author later told a historian that "one of the key misunderstandings in the entire project was the misinterpretation of the word 'fair' as used by the Joint Chiefs. At the time, it never occurred to [him] that the lack of figures might lead to a misunderstanding."[3]

From a rationalist standpoint, this behavior is clearly problematic. If all major military decisions involve making trade-offs, and if analyzing those trade-offs requires assessing probabilities, then national security officials would ideally assess those probabilities as carefully as possible. Yet to say that some practice is problematic does not imply that there are good solutions, and there are several reasons why national security officials argue that they are better off leaving their probability assessments vague, even if this allows for the kind of confusion seen in the Bay of Pigs episode. Many national security analysts believe that making clear probability assessments would warp the psychology of national security decision making by lending illusions of rigor to subjective opinions. Others worry that asking analysts to resolve ambiguous judgments would inappropriately curtail decision makers' freedom of action, or that this could amplify the pressures that analysts face to tell decision makers what they want to hear. I will therefore explain in this chapter how long-standing debates about probability assessment in national security are not simply a matter of semantics. Instead, these debates reflects their own basic trade-offs in the politics, management, and psychology of high-stakes decision making.

There are two main reasons why this subject deserves close scrutiny. First, few issues in public policy analysis are as generalizable as this one. As Ian Hacking put it, "probability ... lies at the basis of all reasonable choices made by officials."[4] While public officials can always leave their probability assessments vague, that does not make these

[3] Wyden (1979, 88–90). [4] Hacking (1990, 4).

judgments any less important. If you support expending resources for any purpose, this means you believe that the chances of achieving positive outcomes are high enough to make that investment worthwhile. To the extent that scholars, practitioners, and pundits are often reluctant to clarify these assumptions, it reveals their discomfort in addressing the logical foundations of high-stakes decision making. And given how assessments of uncertainty appear throughout nearly every intelligence report, military plan, and policy debate, even minor improvements in probabilistic reasoning could have major aggregate impact.

This reluctance to making clear probability assessments also engages broader controversies about how decision makers should combine analysis and intuition when addressing high-stakes issues. Following the work of psychologists such as Daniel Kahneman, Paul Slovic, and Amos Tversky, scholars have identified many ways in which decision makers' intuitions involve misleading heuristics and biases.[5] It is often possible to combat these heuristics and biases through structured reasoning. But structured reasoning has defects, too, and many attempts to overcome the "fog of war" through analytic rigor have ultimately proven to be counterproductive.[6] Because probability assessment in national security decision making is both critically important and exceptionally subjective, it is an area where we might expect both analysis *and* intuition to be flawed. Practical debates about the conduct of probability assessment thus raise deeper questions about what it means to make a sound decision on the basis of subjective judgment.

The following sections explore this controversy in five parts. The first section provides more detail on how national security officials are often reluctant to make clear probability assessments when debating high-stakes issues. I focus this discussion primarily on examples drawn from the US, but similar themes are well documented in Britain, Canada, Israel, and several other countries.[7] The next three sections describe a series of trade-offs involving the psychology, management, and politics of probability assessment in national security. In each area, I describe

[5] On the heuristics and biases research program and its applications to international relations, see Kahneman et al. (1982); Levy (2013); Hafner-Burton et al. (2017).

[6] For debates about the proper balance of analysis versus intuition in national security and other areas of high-stakes decision making, see Gigerenzer (2000); Renshon and Larson (2003); Connable (2012).

[7] See Dhami (2013); Barnes (2016); Lanir and Kahneman (2006).

the extent to which empirical scholarship provides a basis for weighing the costs and benefits of vague information.[8] The chapter's fifth section closes by drawing practical implications from my analysis. In particular, I explain how the goal of improving probability assessment in national security decision making is not just an issue for government officials, but that this is also a question of how scholars, journalists, and pundits can make public discourse as informative as possible.

Words of Estimative Probability

Few scholars or practitioners would challenge the idea that uncertainty surrounds national security decision making. The US Marine Corps's capstone doctrine thus explains that "war is intrinsically unpredictable. At best, we can hope to determine possibilities and probabilities."[9] US Army planning doctrine tells readers that "risk, uncertainty, and chance are inherent in all military operations."[10] The 2005 National Defense Strategy even argued that "uncertainty is the defining characteristic of today's strategic environment."[11]

Yet despite emphasizing the importance of uncertainty in national security decision making, many public officials are reluctant to analyze that subject directly. The seminal essay on this topic is a 1964 article by a high-ranking intelligence official named Sherman Kent, titled "Words of Estimative Probability." In that essay, Kent described leading the production of a National Intelligence Estimate that warned of a "serious possibility" that the Soviet Union would invade Yugoslavia. Kent later asked his analysts to describe what they believed a "serious possibility" entailed. He found that their answers ranged from 20 to 80 percent. Surely intelligence estimates provided little value, Kent argued, if their own authors disagreed so strongly about what their conclusions meant.[12]

In order to address this problem, Kent recommended that intelligence analysts communicate probability assessments using an "odds table," which mapped common qualitative expressions onto numeric percentages. For example, Kent defined the phrase "almost certain" as representing estimates from 87 percent to 99 percent, and he defined

[8] For a review of foundational scholarship on this topic in the decision sciences more broadly, see Wallsten (1990).

[9] US Marine Corps (1997, 7).

[10] US Army (2008, para. 6-100).

[11] White House (2005, 2).

[12] Kent (1964).

the phrase "about even" as representing estimates between 40 percent and 60 percent.[13] Kent's proposal met overwhelming resistance from colleagues who believed that he was asking intelligence analysts to offer judgments with "bogus precision."[14] The CIA thus notes in a preface to an online version of the essay that "Kent's efforts to quantify what were essentially qualitative judgments did not prevail."[15] It took more than forty years after Kent published his article – and controversy over assessments of uncertainty surrounding Iraq's weapons of mass destruction programs, in particular – before intelligence agencies began to implement systematic measures to clarify their probability assessments.[16] Even then, many agencies continued to resist proposals to give these judgments any kind of precise meanings.

For example, the Defense Intelligence Agency's 2015 guidelines for *Expressing Analytic Certainty* define the concept of likelihood itself as "a narrative (rather than numerical) statement." These guidelines give analysts a list of thirty-four approved phrases for expressing probability, while emphasizing that none of these phrases means anything specific: with italicization in the original document, these guidelines state, "*You will note that there are no percentage ranges on the accompanying charts. DIA does not condone the use of probability percentages in its products to portray likelihood.*"[17]

Though controversy over the value of precision in assessing probability comprises one of the oldest debates in intelligence studies, similar questions appear in nearly any other area of national security decision making. For example, US Army doctrine on risk management instructs military planners to divide risks into five categories: frequent, likely,

[13] Kent based these numbers on a survey he conducted that asked analysts and decision makers to say how they might quantify those expressions. In this way, he attempted to define a lexicon that mapped qualitative expressions onto their intuitive definitions.

[14] Though Kent published "Words of Estimative Probability" in 1964, he had circulated classified versions of the essay within the Intelligence Community for a decade. The final version of the essay thus contains Kent's proposal along with his reflections on why that proposal failed.

[15] www.cia.gov/library/center-for-the-study-of-intelligence/csi-publications/books-and-monographs/sherman-kent-and-the-board-of-national-estimates-collected-essays/6words.html. On the history of subsequent attempts to clarify probability estimates in intelligence reporting, see Nye (1994); Weiss (2008); Marchio (2014).

[16] On the history of these reforms, see Fingar (2011); Wheaton (2012).

[17] Defense Intelligence Agency (2015, 1).

occasional, seldom, and unlikely.[18] The definitions of these terms can be difficult to understand. For example, Army doctrine defines a "frequent" risk by saying that, "given 500 or so exposures to the hazard, expect that it will definitely happen to someone." The word "likely" then describes risks such that, "given 1,000 or so exposures without proper controls, it will occur at some point." The trouble here is not just that these definitions lie far from intuitive interpretations of words like "frequent" and "likely." What does it mean to "expect that something will definitely happen"? Is every person in the unit exposed to the risk, so that a "likely" risk should occur at some point if a unit of one thousand soldiers takes an action? Or does the exposure refer to the action itself, so that everyone in the unit could take the action several hundred times and still expect to be safe? Depending on the answer to these questions, interpretations of risk can span huge amounts.

Moreover, problems with vague probability assessments in national security decision making run deeper than debates about linguistic precision. In many cases, national security officials are instructed to assess critical probabilities using relational language that has no clear connection to the number line at all. This practice is most prominent in military planning, where doctrine instructs commanders to select the course of action that minimizes risks and maximizes the chances of success, without necessarily determining what those risks and chances are. For example, US Army doctrine presents the following guidelines for evaluating potential courses of action (COAs):

> The staff compares feasible COAs to identify the one with the highest probability of success . . . The selected COA should also pose the minimum risk to the force and mission accomplishment; place the force in the best posture for future operations; provide maximum latitude for initiative by subordinates; provide the most flexibility to meet unexpected threats and opportunities; provide the most secure and stable environment for civilians in the AO; [and] best facilitate initial information themes and messages.[19]

Of course, military officers should seek to maximize benefits and minimize risks, all else being equal. At the same time, these relational judgments do not imply that a course of action is actually worthwhile. Just because some course of action offers the best chances of success does not mean that those chances are high enough to justify. Meanwhile,

[18] US Army (2006, para. 1-8). [19] US Army (2010, para. B-173).

saying that some course of action offers the lowest risk does not rule out the possibility that this risk is still too high for a rational decision maker to accept.

These concerns are not just relevant to the way that intelligence analysts and military planners assess uncertainty on a regular basis, but they also pertain to debates about national security decision making at the highest levels. For example, when President Barack Obama appointed General Stanley McChrystal to command US forces in Afghanistan in 2009, he asked McChrystal to conduct a strategic review of the problems that the US faced in containing the Taliban insurgency. When General McChrystal provided this assessment, he requested that President Obama adopt a new counterinsurgency strategy bolstered by forty thousand additional troops. General McChrystal argued that these measures would "improve effectiveness" and that they offered "the best prospect for success in this important mission."[20] When this report leaked to the public, it served to build political momentum for what became known as the Afghan Surge.

General McChrystal's report was a candid document, describing many existing flaws with US strategy and explaining that success remained uncertain even with a major influx of resources. Yet it was trivial to argue that deploying forty thousand additional troops would improve effectiveness. The real question was *how much* the Afghan Surge would improve effectiveness, and whether that would raise the probability of strategic success enough to justify the strategy's costs. Meanwhile, the "best prospect for success" in Afghanistan could have meant just about anything. Logically speaking, there was no way to evaluate the merits of General McChrystal's proposal without estimating the chances of success in Afghanistan before and after the recommended deployment. Yet General McChrystal's report left these estimates to the reader.

An even more troubling example of paying limited attention to probability assessment involves President Lyndon Johnson's decision to escalate the Vietnam War in 1965. As with General McChrystal's report on the Afghan Surge, the issue is not so much that top officials estimated their chances of success imprecisely but rather that they largely avoided discussing this subject at all. Thus in their famous book that defends many aspects of Vietnam War decision making, Leslie

[20] COMISAF (2009, 1-3, 1-4, and 2-22).

Gelb and Richard Betts explain that US officials rarely discussed the chances that the measures they proposed would succeed. Instead, Gelb and Betts describe how "debates revolved around how to do things better, and whether they could be done, not whether they were worth doing."[21]

Prior to deploying combat forces to Vietnam, Gelb and Betts write, the Pentagon produced just one major report assessing whether it was feasible for those forces to achieve their intended goals.[22] And even that report avoided addressing this issue in probabilistic terms, concluding that "there appears to be no reason we cannot win if such is our will," while declining to offer any "assessment of the assurance the U.S. can have of winning."[23] Secretary of Defense Robert McNamara later admitted that top officials simply neglected to address this matter, even as they prepared to send hundreds of thousands of US soldiers to fight in Vietnam. In his memoirs, McNamara explained that "we never carefully debated what US force would ultimately be required, what our chances of success would be, or what the political, military, financial, and human costs would be if we provided it. Indeed, these basic questions went unexamined. We were at the beginning of a slide down a tragic and slippery slope."[24]

One of the reasons that many critics find this behavior to be problematic is that it seems so unnecessary. National security officials can always debate probability assessments explicitly if they choose to do so. This would not make their judgments any less subjective, but at least those judgments would then be clear. That would, in turn, mitigate all of the problems described so far in this chapter: confusion about what qualitative language means, questions about what the "best chances of success" or the "lowest risks" entail, or situations in which decision makers simply fail to debate probabilistic assumptions in meaningful ways.[25]

Moreover, despite protestations that probability assessments are "essentially qualitative" or "narrative rather than numerical" (as we have seen the CIA and DIA arguing, respectively), analysts always have a coherent theoretical basis for translating subjective probability

21 Gelb and Betts (1971, 146). 22 Gelb and Betts (1979, 125).
23 Goodpaster Report (1965). 24 McNamara (1995, 107).
25 See, e.g., Schrage (2005); Tetlock and Gardner (2015).

assessments into numbers.[26] Recent research also demonstrates that this precision is empirically meaningful. One study of more than a thousand Canadian intelligence forecasts determined that analysts could reliably divide probability assessments into nine separate tiers.[27] A different study sponsored by the US Intelligence Community collected nearly one million geopolitical forecasts in the form of numeric percentages. Researchers found that coarsening those estimates into "words of estimative probability" consistently degraded their predictive accuracy.[28] Thus the common practice of leaving probability estimates vague does not reflect some inherent epistemic limit on the conduct of national security decision making: rather, it is a choice that appears to sacrifice meaningful information. In order to justify the conventional practice of leaving probability assessments vague, it is thus necessary to explain why making those judgments explicit would somehow harm the quality of national security decision making, and that is the subject to which this chapter now turns.

How Much Attention Should Decision Makers Pay to Subjective Probability Estimates?

One of the main benefits of debating probability estimates explicitly is that this makes it harder for decision makers to overlook important risks. That concern was most clearly displayed by the Bay of Pigs case, in which President Kennedy thought that a "fair chance of success" reflected a supportive assessment instead of a red flag. Similarly, many scholars believe that one of the primary flaws with the US intelligence assessments regarding Iraq's presumed weapon of mass destruction programs was that they did not adequately emphasize the uncertainty surrounding these judgments, thereby allowing readers to develop a false impression of certainty.[29] Even if intelligence analysts had stated that there was a 90 percent chance that Iraq was pursuing nuclear weapons, then no one could have thought that they believed this judgment was certain.

[26] For formal proofs of this argument, see Savage (1954); Ellsberg (1961). For a discussion of this point as applied to intelligence analysis, see Friedman and Zeckhauser (2015).

[27] Mandel and Barnes (2014).

[28] Friedman et al. (forthcoming).

[29] Jervis (2011, Chapter 3).

At the same time, many scholars and practitioners worry that making probability assessments too clear – and especially expressing those judgments quantitatively – could engender the opposite problem, creating illusions of rigor that make subjective judgments appear to be more reliable than they really are. Thus the National Intelligence Council has defended its use of ambiguous language by stating that "assigning precise numerical ratings to such judgments would imply more rigor than we intend."[30] The National Research Council's governing board on risk assessment similarly warns that "while quantitative risk assessment facilitates comparison, such comparison may be illusory or misleading if the use of precise numbers is unjustified."[31]

Concerns over illusions of rigor are especially prominent in scholarship on military planning. Because military operations are so complex, any attempts to predict their potential outcomes are subjective and error-prone.[32] For that reason, many scholars and practitioners believe that it is foolish to place much weight on strategic predictions. Secretary of Defense James Mattis thus argued in an influential 2008 essay that "it is not scientifically possible to accurately predict the outcome of [a military] action. To suggest otherwise runs contrary to historical experience and the nature of war."[33]

These ideas present national security decision makers with a dilemma. At one level, Mattis and other authors are right to say that efforts to predict military outcomes – or to assess other controversial elements of uncertainty in national security decision making – deserve healthy skepticism.[34] At the same time, decision makers cannot do their jobs without assessing these issues. Every national security decision constitutes an implicit prediction about how allocating resources will help to bring about better outcomes.[35] As a result, the key question is not *whether* national security decision makers should assess probability. The real question is *how* they can debate their probabilistic assumptions in a manner that promotes clear thinking without placing undue faith in subjective opinions.

[30] For example, see the guidelines for "Explaining Estimative Language" in the National Intelligence Council's 2007 National Intelligence Estimate on Iran: Nuclear Intentions and Capabilities.
[31] National Research Council (1981, 15).
[32] See Connable (2012) and note 2.
[33] Mattis (2008, 18–19); also Van Riper (2009).
[34] Tetlock (2005).
[35] See Betts (2000); Pape (1997–1998).

Existing research in the decision sciences suggests that decision makers do, in fact, pay closer attention to probability estimates that are expressed at greater levels of precision. For example, Nathan Dieckmann, Paul Slovic, and Ellen Peters conducted a series of experiments in which they presented respondents with sample intelligence reports. Those reports contained probability assessments as well as narrative information. The researchers then randomly varied whether those probability assessments were expressed in qualitative or quantitative form, and they found that respondents who received numeric probability estimates were more likely to rely on that information when interpreting scenarios.[36] Psychologists have replicated these findings in several other contexts where respondents are asked to make gambles depending on a combination of information about the gamble's potential outcomes as well as the probabilities that those outcomes will occur. Generally speaking, these respondents assign greater weight to probability assessments, relative to other cues, when those probability assessments are expressed more precisely.[37]

Yet placing more emphasis on probability assessments is only a bad thing if decision makers already give those judgments proper weight. And international relations scholarship provides grounds for believing that national security officials would be better off giving closer scrutiny to the uncertainties that surround major decisions. For example, Aaron Rapport has shown that the Johnson administration's lack of interest in evaluating its strategic prospects in Vietnam exemplifies a broader tendency among national security decision makers to evaluate long-term objectives based on desirability over feasibility.[38] Dominic Johnson argues that national security officials are generally prone to overconfidence and false certainty, while Yuen-Foon Khong explains that foreign policy decision makers tend to establish expectations of a policy's efficacy by drawing misleading analogies to overly favorable cases.[39] More generally, intelligence scholars often criticize decision makers for demanding overly simplistic, yes-or-no answers to questions that are probabilistic in nature.[40] By making their probability assessments more explicit, national security analysts can combat this desire for certainty.

[36] Dieckmann et al. (2009). [37] Gonzalez-Vallejo et al. (1994); Erev et al. (1993).
[38] Rapport (2015). [39] Johnson (2004); Khong (1992).
[40] See, e.g., Agrell and Treverton (2015).

Moreover, the notion that clear probability estimates would create illusions of rigor rests on an extremely cynical view of national security decision makers' cognitive abilities. The idea must be that these decision makers are so unaware of the subjective nature of these judgments that they will lose sight of that fact simply as a result of the language that analysts use to present their opinions. This claim is not at all obvious, and empirical studies contradict it. Thus, in a recent set of survey experiments, researchers asked military officers to evaluate decisions to use force under conditions of uncertainty. The scenarios varied in whether they conveyed probability estimates in qualitative versus quantitative form. Respondents were then asked whether they supported the use of force, opposed the use of force, or preferred to gather more information before reaching their decision. If quantitative probability assessments created illusions of rigor, then military officers who received those assessments should have been more willing to reach decisions on the basis of the information that they already possessed. But the researchers found that the opposite was true: that military officers who received quantitative probability assessments were more inclined to delay choices until they had gathered more information.[41] That is exactly what one would expect if clearer probability estimates cause decision makers to become more sensitive to the downside risks of military actions, and it is contrary to the expectation that explicit judgments create harmful illusions of rigor.

For these reasons, existing scholarship suggests that the potential benefits of encouraging national security decision makers to pay more careful attention to key risks outweigh the potential costs of surrounding subjective judgments with illusions of rigor. However, most of the existing psychological scholarship on this issue examines issues outside of national security decision making, and even when scholars conduct experimental research on elite samples, those experiments cannot possibly capture the complexity of actual, high-stakes decisions.[42] It is thus most accurate to say that the "illusions of rigor" concern, while plausible, currently lacks a rigorous evidentiary basis, and therefore represents a weak justification for leaving key judgments deliberately vague.

[41] Friedman et al. (forthcoming). The authors of this study also found that the military officers were just as "confident" in making these choices: this is to say that they did not request more information simply because they were confused by receiving probability estimates in quantitative form.

[42] Hafner-Burton et al. (2013).

Whose Job Is It to Resolve Ambiguous Judgments?

A second set of trade-offs relating to the conduct of probability assessment in national security decision making involves the question of who holds the responsibility for interpreting ambiguous issues. For the purposes of this discussion, I will refer to individuals who have the authority to make a particular choice as "decision makers," while using the term "analysts" to describe individuals who provide information that *helps* decision makers to evaluate those choices. In this sense, the distinction between who is an analyst and who is a decision maker can vary by context: thus the Secretary of Defense is the top decision maker in managing the Pentagon, but an analyst when helping the president determine whether or not to go to war.

Decision makers must delegate many key elements of national security policy. For example, no one expects the president of the United States to craft every detail of a given military strategy. But when it comes to judging important, subjective issues, such as estimating the chances that a military strategy will succeed, one can raise legitimate questions about whose judgment matters most. In particular, some scholars and practitioners worry that if analysts resolve this ambiguity on their own, then this would unduly constrain decision makers' freedom of action.

The clearest examples of this problem occur in regulatory policy. Regulatory bodies such as the Environmental Protection Agency, the Food and Drug Administration, and the Department of Transportation are legally required to subject proposed regulations to cost–benefit analyses. In conducting these analyses, agencies must determine how they will value the outcomes that regulations aim to achieve. (For example, most US regulatory agencies currently estimate the value of saving a life at roughly ten million dollars.) Analysts then estimate a probability distribution of expected benefits that the regulation might generate, such as the number of deaths that a traffic safety measure might prevent. By multiplying these probability distributions and outcome valuations together, a regulatory agency can determine whether or not the expected benefits of a proposed regulation exceed its expected costs.[43]

Since these judgments are usually subjective, reasonable people can disagree about what they should entail. But once analysts make these judgments explicit, this leaves decision makers with limited discretion.

[43] For a description of this process, see Sunstein (2014).

When outcome measures like the value of a statistical life are generally established in advance, and when a regulation's costs are often relatively straightforward to calculate, quantifying a probability distribution of potential benefits is largely equivalent to answering the question of whether or not a regulation is cost-effective. In some contexts, scholars and practitioners worry that this is inappropriate, because it means that analysts are effectively making high-stakes decisions themselves.[44]

A similar example within the context of national security decision making involves the assessment of uncertainty surrounding drone strikes. According to the Obama administration's guidelines for approving drone strikes, counterterrorism officials can only use lethal force against high-value targets when there is "near certainty" that they have correctly identified their target and "near certainty" that the strike will not harm noncombatants.[45] Of course, this raises the question of what it means to be "nearly certain" in judging those issues. One might argue that it is inappropriate for analysts to make that determination, on the grounds that doing so would constrain decision makers' autonomy for approving lethal action.[46]

Yet giving decision makers the option to resolve ambiguity however they see fit raises concerns of its own. Generally speaking, decision makers possess limited time and proficiency for processing difficult information. Even more concerning is the notion that decision makers may be inclined to interpret vague information in a biased fashion that reinforces their prior beliefs. As Richard Betts put it, "the greater the ambiguity, the greater the impact of preconceptions."[47]

A substantial volume of international relations scholarship substantiates this concern. For example, Keren Yarhi-Milo has demonstrated that many national security decision makers overrate their abilities to assess uncertainty, and that their perceptions of ambiguous issues are often driven by gut instincts or extraneous cues rather than

[44] Wallsten (1990, 34–35); Lave (1987). [45] Obama (2013).

[46] An analogy to this concern is the question of what it means for a criminal defendant to be "guilty beyond a reasonable doubt." US courts leave this standard deliberately vague, on the grounds that one of a jury's primary responsibilities is to establish that standard on a case-specific basis. However, critics of this practice worry that this vagueness adds an unnecessary degree of arbitrary judgment to the application of criminal justice. For more on this controversy, see Tillers and Gottfried (2006).

[47] Betts (1982, 103).

careful assessments of available evidence.[48] Robert Jervis and Philip Tetlock have argued that decision makers have natural cognitive impulses to interpret ambiguous information in self-serving ways.[49] Joshua Rovner has shown that this tendency is not just the result of psychological distortions, but that decision makers sometimes deliberately manipulate ambiguous intelligence in order to support their policy agendas.[50] Some foreign policy officials believe that such behavior is simply part-and-parcel of the democratic process: thus former US Secretary of State Dean Acheson famously wrote in his memoirs that it was a public officer's job to make ambiguous issues seem "clearer than truth" in order win political arguments.[51]

The question of who bears professional responsibility for resolving ambiguous judgments thus presents a second key trade-off in handling high-stakes probability assessments. When analysts resolve this ambiguity they may constrain decision makers' freedom of action; but when decision makers resolve this ambiguity, they might do so in a manner that is overly favorable to their prior beliefs. How should national security officials manage this trade-off?

In answering this question, it is important to consider just how much it would constrain decision makers' freedom of action if national security analysts provided clearer assessments of uncertainty than they do now. The Obama administration's drone strike guidance offers a concrete example of this problem. But in most other cases, that concern appear to be relatively small. The reason why probability assessments are so controversial in regulatory affairs is that monetary cost–benefit analyses involve clear break-even points for determining which actions create positive expected value. Virtually all high-stakes national security decisions lack such analytic clarity. In most cases, it is a highly subjective matter as to what the national interest entails, let alone how decision makers could combine all elements of the national interest into a single index that they could quantify in monetary terms. This means that there is usually no clear answer as to what level of certainty would be sufficient to justify a national security decision. By extension, there is usually little reason to believe that asking analysts to

48 Yarhi-Milo (2014).

49 Tetlock (2005); Jervis (1976). On how these cognitive mechanisms allow decision makers to maintain an illusion of efficacy in handling complex issues, see Dawes (1988).

50 Rovner (2011).

51 Acheson (1969, 375).

provide clear assessments of uncertainty would curtail decision makers' options.

For example, consider what might have happened if General McChrystal had told President Obama that the Afghan Surge had a 40 percent chance of success instead of arguing that it represented "the best prospect" that was presently available. How would that have forced the president's hand in determining whether or not that strategy was in the country's best interest? Depending on how President Obama valued the prospective costs and benefits that the Afghan Surge might entail, he could have decided that a 40 percent chance was too low to be worth the risk. But President Obama could also have decided that the stakes in Afghanistan were great enough to justify the gamble. Since there was no clear bar for how high the probability of success in Afghanistan needed to be in order to justify the Afghan Surge, making this estimate explicit would have given President Obama a clearer sense of how General McChrystal viewed this strategy's prospects without closing off his options for determining what those prospects meant.

Concerns about restricting the president's autonomy by providing clear probability assessments seem particularly out of place in this example given that General McChrystal was not presenting his assessment impartially. General McChrystal wrote his report with the goal of convincing President Obama to approve the Afghan Surge.[52] In order to believe that this recommendation was worthwhile, General McChrystal had to assume that the Afghan Surge's chances of success were high enough to make the strategy's expected benefits exceed its expected costs. To say that it would have been inappropriate for General McChrystal to assess those chances of success directly is thus to argue that commanders should be able to make recommendations to the president without explaining the assumptions on which they base those recommendations.

If anything, clearly assessing the Afghan Surge's probability of success might have given President Obama *more* freedom of action in this case. Several accounts of this episode indicate that President Obama felt that military advisers left him with limited options, especially after General McChrystal's report was leaked to the public.[53] Had that report indicated that the Afghan Surge stood a 40 percent chance of

[52] McChrystal (2013, Chapter 19).

[53] On the Afghan Surge debate, see Marsh (2014); Woodward (2010); Davidson (2013).

success – or had it provided any other estimate that fell substantially short of certainty – President Obama might have possessed a greater opportunity to say that these prospects simply seemed too small to be worth the lives and resources that the Afghan Surge would consume.

On balance, this appears to be another area in which the trade-offs fall in favor of encouraging national security officials to make clearer probability assessments. Existing scholarship provides rigorous, empirical documentation of how cognitive biases and political incentives can cause decision makers to misinterpret ambiguous judgments. At the same time, most national security decisions do not contain the kind of dynamics seen in regulatory policy, in which resolving ambiguous assessments of uncertainty can directly constrain decision makers' freedom of action.

Accounting for Accountability

One reason why many scholars believe that vague probability assessments are problematic is because this vagueness impedes accountability. If an analyst says that some outcome is 80 percent likely to occur, then it will become clear after the fact just how right or wrong her estimate was. But if the analyst says that this outcome is "likely" – or if she argues, even more vaguely, that this outcome is "possible" – then there is room for interpretation in evaluating the quality of her judgment.[54] Philip Tetlock argues that this evasion of accountability is a consistent problem in national security discourse. "Both private- and public-sector prognosticators must master the same tightrope-walking act," he writes.

> They know they need to sound as though they are offering bold, fresh insights into the future not readily available off the street. And they know they cannot afford to be linked to flat-out mistakes. Accordingly they have to appear to be going out on a limb without actually going out on one. That is why [foreign policy analysts] so uniformly appear to dislike affixing "artificially precise" subjective probability estimates to possible outcomes... It is much safer to retreat into the vague language of possibilities and plausibilities.[55]

[54] Psychologists call this behavior "elastic redefinition." See Piercey (2009). Hood (2011) explains how public officials often abstain from assessing controversial issues to minimize their exposure to blame.

[55] Tetlock (2009, 67).

The difficulty of evaluating vague probability assessments has two perverse effects. First, this makes it hard to know how much faith readers should place in analytic products. Without a clear track record of performance, it is hard to gauge the quality of national security analysis, to develop rigorous impressions of what kinds of issues analysts do better at assessing than others, or even to say whether intelligence officials and military planners provide more accurate judgments than columnists in the *New York Times*. Second – and perhaps of greater concern – is that the harder it becomes to hold national security analysts accountable for the accuracy of their judgments, the less incentive those analysts possess to make accurate judgments in the first place.[56]

However, existing scholarship suggests that there are two relevant conditions under which efforts to hold analysts accountable for their judgments can backfire.[57] The first of these conditions is when analysts report to decision makers who have known views. In this context, many people display tendencies to present their judgments in a manner that will be more favorable to their audience.[58] In some cases, this might reflect a deliberate attempt to curry favor or to avoid criticism. But it also appears that most people naturally prefer consensus to disagreement. Thus Irving Janis famously argued in his book *Victims of Groupthink* that national security officials pay social and psychological costs for breaking consensus, and that they will therefore alter or self-censor dissenting viewpoints in order to avoid that discomfort.[59] If this is true, then it means that asking analysts to make their judgments clearer could end up creating the impression that group consensus is stronger than it really is.

A second condition under which efforts at promoting accountability can have negative consequences is when evaluations of performance are either subjective or delayed.[60] Under these conditions, decision makers have greater flexibility to sanction subordinates who provide unfavorable information. This dynamic is especially well documented in the context of US decision making with respect to Vietnam, where

[56] For a review of debates about evaluating the quality of intelligence analysis, specifically, see Tetlock and Mellers (2011); Friedman and Zeckhauser (2016).
[57] For a review of scholarship on accountability, see Lerner and Tetlock (1999).
[58] For examples, see Lerner and Tetlock (1999, 256–257).
[59] Janis (1972). See also Schafer and Crichlow (2010); Mintz and Wayne (2016).
[60] Prendergast (1993).

President Johnson developed a reputation for punishing advisers who disagreed with his policies. For example, after Vice President Hubert Humphrey wrote Johnson a memorandum that carefully articulated his objections to escalating the war, Johnson cut Humphrey out of high-level national security discussions. Several historians thus believe that one reason why other top officials preferred to avoid assessing their strategic prospects in Vietnam directly was because this would force them into a dilemma of either misrepresenting their beliefs or losing power.[61] Even though Humphrey and other skeptics were vindicated in the long run, that was well past the point at which they could recover their influence. For these reasons, skeptics might plausibly worry that asking national security officials to make clear probability assessments would only reinforce pressure to tell decision makers what they want to hear.

The notion that efforts to improve accountability can create both favorable and unfavorable incentives for national security analysts thus represents a third set of trade-offs surrounding the conduct of probability assessment in national security decision making. Of the trade-offs described in this chapter, the countervailing effects of accountability are perhaps the hardest to predict. Part of the problem is that there is rigorous empirical research supporting both the positive and negative effects of promoting accountability in such situations. But another obstacle to confronting this dilemma is that most scientific research on accountability, which largely takes place in experimental settings, cannot possibly replicate the density and the magnitude of the pressures that national security analysts face when debating sensitive issues. For that reason, trade-offs involving the politics of accountability arguably present the most challenging set of issues surrounding the conduct of probability assessment in national security decision making.

At the same time, it is important to note that analysts face these trade-offs regardless of how clearly they express their judgments. Communicating uncertainty using "words of estimative probability" spectrums presents the same dilemmas of speaking truth to power. And in many cases, leaving probability assessments vague may only expose national security officials to additional blame, as critics can exploit this ambiguity to make analysts' judgments seem more mistaken than

[61] See, e.g., Logevall (1999); Preston (2006).

they really were.[62] For example, President Kennedy did not give the Joint Chiefs of Staff credit for their foresight in stating that the Bay of Pigs invasion had a "fair chance of success." Even if the Joint Chiefs intended for this ambiguous judgment to serve as a warning, President Kennedy later complained to a colleague that "those sons of bitches" had "just sat there nodding, saying it would work."[63] In this case, offering a clearer probability estimate would surely have given the Joint Chiefs of Staff a better position to defend their judgments after the fact.

More generally, national security analysts would be in a better position to defend themselves against criticism of their performance if they tracked that performance systematically. Consider what happens when intelligence analysts predict that some outcome has an 80 percent chance of taking place. If this estimate is well calibrated, then one time out of five, that outcome will not in fact take place and critics might say that they misjudged the issue. If the analysts are then able to document that outcomes they predict with 80 percent probability actually occur roughly 80 percent of the time, then they would possess a rigorous basis for arguing that this negative surprise reflected the inherent uncertainty of world politics as opposed to flawed judgment. But because most national security agencies do not currently track their performance in such a systematic way, analysts have no such recourse for defending themselves when called to account. Thus while bureaucracies often view systems of accountability as constituting threats to their status, rigorously tracking performance might also help national security agencies to defend themselves against critics who cherry-pick analytic failures without giving corresponding credit for analytic successes.[64]

[62] This is particularly true for the idea that the Intelligence Community claimed it was a "slam dunk" that Iraq was pursuing nuclear weapons. The National Intelligence Estimate on Iraq's weapons programs does not actually contain a single statement assessing the probability that Saddam Hussein was using nuclear weapons, and the "slam dunk" term was used by Director of Central Intelligence George Tenet when telling President Bush that the White House could convince the public that Saddam was a danger – a point on which Tenet later proved to be correct.

[63] Reeves (1993, 103).

[64] On the potential benefits of tracking performance in this way, see Tetlock and Mellers (2011); Friedman and Zeckhauser (2016). Mandel and Barnes (2014) provides evidence that intelligence analysts' performance is much better than

Practical Implications

I began this chapter by explaining how, even though all high-stakes national security decisions are surrounded by uncertainty, national security officials often leave their assessments of that uncertainty deliberately vague. This is most clearly observed through the way in which national security organizations direct intelligence analysts and military planners to express probability using qualitative terms which often lack clear definitions. We have also seen several high-profile cases in which decision makers authorized the use of force with only cursory efforts to scrutinize the chances that those decisions would succeed. I argued that this behavior raises basic problems with the conduct of high-stakes decision making. It is logically impossible to evaluate a policy without making probabilistic assumptions. Why would anyone want to leave those assumptions unclear?

In addressing this issue, I described how national security scholars and practitioners have raised several objections to making clear probability assessments, and especially to debating those assessments quantitatively. Clearer probability assessments would make it harder for decision makers to overlook the uncertainty that surrounds major choices, but this might also create harmful illusions of rigor. Even if analysts are generally in a better position to resolve ambiguity, doing so might constrain decision makers' freedom of action. And efforts to make national security officials more accountable for their judgments can potentially backfire by giving analysts incentives to tell decision makers what they want to hear. In each of these ways, I argued that controversy over the manner in which national security officials debate uncertainty runs deeper than semantics. Instead, this controversy highlights basic trade-offs with the psychology, the management, and the politics of national security decision making.

In addition to describing these trade-offs, this chapter explained where existing scholarship stands on each issue. Because many of these trade-offs remain open to legitimate disagreement, I will refrain from offering an overall judgment about the relative merits of "words of estimative probability," numeric percentages, or any other methods for debating probability assessments in national security decision making.

what public opinion assumes. On the notion that intelligence analysts are unfairly held to account for their failures instead of their successes, see Betts (2006).

However, I believe that this review suggests two main practical takeaways.

First, when national security officials debate major policies, it is important to assess the chances that those policies will succeed in absolute terms. Assessing strategic prospects using relational judgments – that is, describing what measures offer the "best" chances of success or the "lowest" risks – is an inherently misleading way to debate high-stakes choices.[65] The notion that military doctrine encourages commanders to select and to justify courses of action on that basis is unambiguously problematic. Decision makers should be skeptical of these arguments. Any time an adviser describes a policy as presenting the "best prospect" or argues that it will "improve" the chances of success, it is important to ask exactly what these prospects or improvements might entail. If the adviser cannot provide a meaningful answer to this question, then she is not in a position to be recommending that policy to begin with.

Second, and more broadly, even though the examples in this chapter have mainly revolved around the behavior of public officials, questions about probability assessment in national security decision making are also relevant to structuring the broader public discourse. When scholars and pundits debate the merits of major policies, they too must wrestle with uncertainty. In doing so, their opinions can be just as vague and as uninformative as the examples that this chapter has provided.[66] And that vagueness can be even more pernicious. When intelligence analysts write reports, it is at least possible that decision makers will know how to reference key terms against official definitions. By contrast, when a pundit says that some outcome "might" happen, or argues that some risk is "likely" to occur, readers have little recourse for understanding what these vague terms mean. For this reason, the arguments that this chapter has considered regarding the costs and benefits of vague information are not just a matter for evaluating the doctrine employed by intelligence analysts and military planners. Instead, these arguments raise broader questions about what levels of clarity and rigor the public should demand when debating high-stakes issues.

[65] Thus pharmaceutical companies routinely goad customers into purchasing treatments that promise to reduce risks, even if those risks are already vanishingly small. See Gigerenzer (2002).

[66] Tetlock and Gardner (2015); Silver (2012).

References

Acheson, Dean. 1969. *Present at the Creation: My Years in the State Department*. New York: W. W. Norton.

Agrell, Wilhelm, and Gregory Treverton. 2015. *Intelligence and Science*. New York: Oxford University Press.

Barnes, Alan. 2016. Making Intelligence Analysis More Intelligent: Using Numeric Probabilities. *Intelligence and National Security* 31: 327–344.

Betts, Richard K. 1982. *Surprise Attack*. Washington, DC: Brookings Institution Press.

2000. Is Strategy an Illusion? *International Security* 25: 5–50.

2006. *Enemies of Intelligence: Knowledge and Power in American National Security*. New York: Columbia University Press.

Beyerchen, Alan. 1993–1993. Clausewitz, Nonlinearity, and the Unpredictability of War. *International Security* 17: 59–90.

Beyth-Merom, Ruth. 1982. How Probable Is Probable? A Numerical Translation of Verbal Probability Expressions. *Journal of Forecasting* 1: 257–269.

Clausewitz, Carl von. 1976. *On War*. Trans. Michael Howard and Peter Paret. Princeton, NJ: Princeton University Press.

COMISAF [Commander, International Security Assistance Force-Afghanistan]. 2009. *COMISAF Initial Assessment*. Kabul: Headquarters ISAF.

Connable, Ben. 2012. *Embracing the Fog of War: Assessment and Metrics in Counterinsurgency*. Santa Monica, CA: RAND Corporation.

Davidson, Janine. 2013. Civil-Military Friction and Presidential Decision Making: Explaining the Broken Dialogue. *Presidential Studies Quarterly* 1: 129–145.

Dawes, Robyn M. 1988. *Rational Choice in an Uncertain World*. San Diego, CA: Harcourt Brace Javonovich.

Defense Intelligence Agency. 2015. Expressing Analytic Uncertainty. Tradecraft Note 01-15. January.

Dhami, Mandeep K. 2013. *Understanding and Communicating Uncertainty in Intelligence Analysis*. London: Report Prepared for H.M. Government.

Dieckmann, Nathan F., Paul Slovic, and Ellen M. Peters. 2009. The Use of Narrative Evidence and Explicit Likelihood by Decisionmakers Varying in Numeracy. *Risk Analysis* 29: 1473–1488.

Ellsberg, Daniel. 1961. Risk, Ambiguity, and the Savage Axioms. *Quarterly Journal of Economics* 75: 643–669.

Erev, Ido, Gary Bornstein, and Thomas S. Wallsten. 1993. The Negative Effect of Probability Assessments on Decision Quality. *Organizational Behavior and Human Decision Making* 55: 78–94.

Fingar, Thomas. 2011. *Reducing Uncertainty: Intelligence Analysis and National Security*. Stanford, CA: Stanford Security Studies.

Friedman, Jeffrey A., Joshua D. Baker, Barbara A. Mellers, Philip E. Tetlock, and Richard Zeckhauser. Forthcoming. The Value of Precision in Probability Assessment: Evidence from a Large-Scale Geopolitical Forecasting Tournament. *International Studies Quarterly*.

Friedman, Jeffrey A., Jennifer S. Lerner, and Richard Zeckhauser. Forthcoming. Behavioral Consequences of Probabilistic Precision: Experimental Evidence from National Security Officials. *International Organization*.

Friedman, Jeffrey A., and Richard Zeckhauser. 2012. Assessing Uncertainty in Intelligence. *Intelligence and National Security* 27: 824–847.

2015. Handling and Mishandling Estimative Probability: Likelihood, Confidence, and the Search for Bin Laden. *Intelligence and National Security* 30: 77–99.

2016. Why Evaluating Estimative Accuracy Is Feasible and Desirable. *Intelligence and National Security* 31: 178–200.

Gelb, Leslie H., and Richard Betts. 1971. Vietnam: The System Worked. *Foreign Policy* 3: 140–167.

1979. *The Irony of Vietnam: The System Worked*. Washington, DC: Brookings Institution Press.

Gigerenzer, Gerd. 2000. *Adaptive Thinking: Rationality in the Real World*. New York: Oxford University Press.

2002. *Calculated Risks: How to Know When Numbers Deceive You*. New York: Simon and Schuster.

Gonzalez-Vallejo, Claudia C., Ido Erev, and Thomas S. Wallsten. 1994. Do Decision Quality and Preference Order Depend on Whether Probabilities Are Verbal or Numerical? *American Journal of Psychology* 107: 157–172.

Goodpaster Report. 1965. *Foreign Relations of the United States, 1964–1968*. Vol. III, Doc. 69.

Hacking, Ian. 1990. *The Taming of Chance*. New York: Cambridge University Press.

Hafner-Burton, Emilie M., Stephan Haggard, David A. Lake, and David G. Victor. 2017. The Behavioral Revolution and the Study of International Relations. *International Organization* 71: S1–S31.

Hafner-Burton, Emilie M., Alex Hughes, and David G. Victor. 2013. The Cognitive Revolution and the Political Psychology of Elite Decision Making. *Perspectives on Politics* 11: 368–386.

Hood, Christopher. 2011. *The Blame Game: Spin, Bureaucracy, and Self-Preservation in Government*. Princeton, NJ: Princeton University Press.

Hyde, Susan D. 2015. Experiments in International Relations: Lab, Survey, and Field. *Annual Review of Political Science* 18: 403–424.

Janis, Irving. 1972. *Victims of Groupthink*. Boston: Houghton Mifflin.
Jervis, Robert. 1976. *Perception and Misperception in International Politics*. Princeton, NJ: Princeton University Press.
1997. *System Effects: Complexity in Political and Social Life*. Princeton, NJ: Princeton University Press.
2011. *Why Intelligence Fails*. Ithaca, NY: Cornell University Press.
Johnson, Dominic D. P. 2004. *Overconfidence in War: The Havoc and Glory of Positive Illusions*. Cambridge, MA: Harvard University Press.
Kahneman, Daniel, Paul Slovic, and Amos Tversky, eds. 1982. *Judgment under Uncertainty: Heuristics and Biases*. New York: Cambridge University Press.
Kent, Sherman. 1964. Words of Estimative Probability. *Studies in Intelligence* 8: 49–65.
Khong, Yuen-Foon. 1992. *Analogies at War*. Princeton, NJ: Princeton University Press.
Lanir, Zvi, and Daniel Kahneman. 2006. An Experiment in Decision Analysis in Israel in 1975. *Studies in Intelligence* 50. www.cia.gov/library/center-for-the-study-of-intelligence/csi-publications/csi-studies/studies/vol50no4/an-experiment-in-decision-analysis-in-israel-in-1975.html.
Lave, Lester B. 1987. Health and Safety Risk Analyses: Information for Better Decisions. *Science* 236: 291–295.
Lerner, Jennifer S., and Philip E. Tetlock. 1999. Accounting for the Effects of Accountability. *Psychological Bulletin* 125: 255–275.
Levy, Jack S. 2013. Psychology and Foreign Policy Decision-Making. In Leonie Huddy, David O. Sears, and Jack S. Levy eds., *The Oxford Handbook of Political Psychology*, 2nd ed., 301–333. Oxford: Oxford University Press.
Logevall, Fredrik. 1999. *Choosing War*. Berkeley: University of California Press.
Mandel, David R., and Alan Barnes. 2014. Accuracy of Forecasts in Strategic Intelligence. *Proceedings of the National Academy of Sciences* 111: 10984–10989.
Marchio, James. 2014. "If the Weatherman Can . . . ": The Intelligence Community's Struggle to Express Analytic Uncertainty in the 1970s. *Studies in Intelligence* 58: 31–42.
Marsh, Kevin. 2014. Obama's Surge: A Bureaucratic Politics Analysis of the Decision to Order a Troop Surge in the Afghanistan War. *Foreign Policy Analysis* 10: 265–288.
Mattis, James N. 2008. USFJCOM Commander's Guidance for Effects-Based Operations. *Parameters* 38: 18–25.
McChrystal, Stanley. 2013. *My Share of the Task*. New York: Penguin.

McNamara, Robert S., with Brian VanDeMark. 1995. *In Retrospect: The Tragedy and Lessons of Vietnam*. New York: Times Books.

Mintz, Alex, and Carly Wayne. 2016. *The Polythink Syndrome*. Stanford, CA: Stanford University Press.

Mosteller, Frederick, and Cleo Youtz. 1990. Quantifying Probabilistic Expressions. *Statistical Science* 5: 2–12.

National Intelligence Council. 2007. *National Intelligence Estimate, Iran: Nuclear Intentions and Capabilities*. www.dni.gov/files/documents/Newsroom/Press%20Releases/2007%20Press%20Releases/20071203_release.pdf.

National Research Council. 1981. *The Handling of Risk Assessments in NRC Reports*. Washington, DC: National Academy Press.

Nye, Joseph S., Jr. 1994. Peering into the Future. *Foreign Affairs* 73: 82–93.

Obama, Barack. 2013. Procedures for Approving Direct Action against Terrorist Targets Located Outside the United States and Areas of Active Hostilities. Presidential Policy Guidance, May 22.

Pape, Robert A. 1997–1998. The Air Force Strikes Back. *Security Studies* 7: 191–214.

Piercey, David M. 2009. Motivated Reasoning and Verbal vs. Numerical Probability Assessment: Evidence from an Accounting Context. *Organizational Behavior and Human Decision Processes* 108: 330–341.

Prendergast, Canice. 1993. A Theory of "Yes Men." *American Economic Review* 83: 757–770.

Preston, Andrew. 2006. *McGeorge Bundy, the NSC, and Vietnam*. Cambridge, MA: Harvard University Press.

Rapport, Aaron. 2015. *Waging War, Planning Peace: US Noncombat Operations and Major Wars*. Ithaca, NY: Cornell University Press.

Reeves, Richard. 1993. *President Kennedy: Profile of Power*. New York: Simon and Schuster.

Renshon, Stanley A., and Deborah Welch Larson, eds. 2003. *Good Judgment in Foreign Policy*. Lanham, MD: Rowman and Littlefield.

Rovner, Joshua. 2011. *Fixing the Facts: National Security and the Politics of Intelligence*. Ithaca, NY: Cornell University Press.

Savage, Leonard J. 1954. *The Foundations of Statistics*. New York: John Wiley.

Schafer, Mark, and Scott Crichlow. 2010. *Groupthink versus High-Quality Decision Making in International Relations*. New York: Columbia University Press.

Schrage, Michael. 2005. What Percent Is Slam Dunk? *Washington Post*, February 20.

Silver, Nate. 2012. *The Signal and the Noise*. New York: Penguin.

Sunstein, Cass R. 2014. *Valuing Life: Humanizing the Regulatory State*. Chicago: University of Chicago Press.

Tetlock, Philip E. 2005. *Expert Political Judgment: How Good Is It? How Do We Know?* Princeton, NJ: Princeton University Press.

2009. Reading Tarot on K Street. *The National Interest* 103: 57–67.

Tetlock, Philip E., and Daniel Gardner. 2015. *Superforecasting: The Art and Science of Prediction*. New York: Crown.

Tetlock, Philip E., and Barbara A. Mellers. 2011. Intelligent Management of Intelligence Agencies: Beyond Accountability Ping-Pong. *American Psychologist* 66: 542–554.

Tillers, Peter, and Jonathan Gottfried. 2006. Case Comment – United States v. Copeland, 369 F. Supp. 2d 275 (E.D.N.Y. 2005): A Collateral Attack on the Legal Maxim That Proof beyond a Reasonable Doubt Is Unquantifiable? *Law, Probability, and Risk* 5: 135–157.

US Army. 2006. Composite Risk Management. Field Manual 5-19.

US Army. 2008. Operations. Field Manual 3-0.

US Army. 2010. The Operations Process. Field Manual 5-0.

US Marine Corps. 1997. Warfighting. Doctrinal Publication 1.

Van Riper, Paul K. 2009. EBO: There Was No Baby in the Bathwater. *Joint Forces Quarterly* 52: 82–85.

Wallsten, Thomas. 1990. The Costs and Benefits of Vague Information. In Robin M. Hogarth, ed., *Insights in Decision Making: A Tribute to Hillel J. Einhorn*, 28-43. Chicago: University of Chicago Press.

Watts, Barry D. 2004. *Clausewitzian Friction and Future War*. Rev. ed. Washington, DC: National Defense University.

Weiss, Charles. 2008. Communicating Uncertainty in Intelligence and Other Professions. *International Journal of Intelligence and CounterIntelligence* 1: 57–85.

Wheaton, Kristan J. 2012. The Revolution Begins on Page Five: The Changing Nature of NIEs. *International Journal of Intelligence and Counter-Intelligence* 25: 330–349.

White House. 2005. *National Defense Strategy of the United States of America*. Washington, DC: White House.

Woodward, Bob. 2010. *Obama's Wars*. New York: Simon and Schuster.

Wyden, Peter H. 1979. *Bay of Pigs: The Untold Story*. New York: Simon and Schuster.

Yarhi-Milo, Keren. 2014. *Knowing the Adversary: Leaders, Intelligence, and Assessment of Intentions in International Relations*. Princeton, NJ: Princeton University Press.

9 *Conclusion*

The Complexity of Military Effectiveness

FILIPPO ANDREATTA

The Limits of Reductionist Theories

Throughout history the problem of military effectiveness has been one of the central questions of international politics.[1] Because the issue of who is likely to prevail in war greatly influences the fate of powers great and small and shapes local and global equilibria, scholars and policy makers alike have long pondered on the roots of military power. Yet, despite a wide and growing body of literature and a lively and sometimes heated debate, there is no consensus in sight on its sources. One of the main reasons for this lack of agreement is the fact that much, if not most, of the literature has tried to find *the* single factor which contributes the most to military effectiveness. As a consequence, many theories in this area of inquiry tend to revolve around a single, major cause for explaining war outcomes. Such an approach may be termed reductionist in the sense that theories "reduce" the overall outcome to a single property/variable of the actors. No doubt, this parsimonious and elegant approach has the quality of simplicity and may be the only one available for scholars who try to formulate broad generalizations across time and space. It also tempts policy makers to look for a simple way to cope with a variegated and worrisome issue, giving them the illusion that there exists a "magic wand" capable of solving the puzzling question of military power. However, as the chapters in this volume seek to demonstrate, such a monocausal path may seldom be adequate. In what has become a classic essay, Isaiah Berlin taught us that the pluralism of the fox outweighs the monism of the hedgehog when it comes to explaining human affairs.[2] Many complex social phenomena, like military conflict, are often just too complicated

[1] Military effectiveness is defined in the introductory chapter as "the degree to which militaries can accomplish at acceptable costs the goals assigned to them by political leaders."
[2] Berlin (1953).

Table 9.1 *Typology of Reductionist Theories of Military Effectiveness*

	Material	Ideational
Military	Technology, quantity of units or equipment	Doctrine, discipline, military leadership
Societal	Industrial power, regime type	Motivation

to be understood by monocausal explanations. In particular, this type of theories display three main shortcomings: the limits of the research design, the propensity to underestimate trade-offs, and a systematic overlooking of the strategic dimension. This book – the outcome of a collaborative international project – seeks therefore to follow a different approach.[3]

The first limit of reductionist theories is that their parsimonious research design hinders their comparative analysis. The literature on military effectiveness has indeed uncovered many possible monocausal explanations,[4] but these are unfortunately incompatible with each other because they relate to different domains. As a crude generalization, as in Table 9.1, monocausal explanations can broadly be classified into four groups according to whether they refer to the strictly military field, rather than to society at large, and whether they deal with material, rather than ideational, variables. First, there are military-material explanations such as those emphasizing the role of technological advance, or quantitative superiority in terms of manpower or certain types of equipment (i.e., the bean counting approach).[5] Second, there are military-ideational explanations based on such factors as the adequacy of military (strategic or tactical) doctrine, the discipline of soldiers, or the quality of military leadership.[6] These explanations have gained in popularity, especially in the last decades, when numerically inferior armies – from Israel in 1967 to the US in 2003 – have

[3] This book is the outcome of a research project coordinated by Dan Reiter and financed by the Trento-based Bruno Kessler Foundation and its Research Center on International Politics and Conflict Resolution (Fbk-Cerpic). The Center fosters the creation international networks of scholars studying international relations with the aim of advancing the development of the discipline in Italy.

[4] Brooks and Stanley (2007); Knox and Murray (2001).

[5] Krepinevich (1994); Horowitz (2010); Rosen (1991); Van Creveld (2010).

[6] Biddle (2004); Cohen (2002); Johnston (1995); Pape (1996); Shils and Janowitz (1948); Snyder (1984); Stam (1996).

gained overwhelming victories on the battlefield. Third, there are societal-material explanations pointing to such factors as regime type (with all its institutional consequences like civil–military relations or the quality of intelligence due to the marketplace of ideas) or industrial power.[7] These explanations stem from the classic idea that the side with the last "thaler" or with the last "escudo" will prevail.[8]

Fourth, there are societal-ideational explanations based on the motivation of the population depending on factors such as nationalism, ideology or the "identity" – social, religious, or ethnic – cleavages within society.[9] While all these theories are plausible and some of them convincing, they cannot be valid at the same time since they are concerned with diverse and incomparable domains. Furthermore, quantitative approaches cannot readily establish which theory exerts more explanatory power, due to notorious difficulties in coding some of these factors (especially the immaterial ones) and in establishing strong correlations with a limited number of (often heterogeneous) observations. In these circumstances, different research designs may lead to opposite conclusions.

Second, as the essays in this volume make abundantly clear, many measures aimed at enhancing military effectiveness involve a trade-off, a situation in which it is necessary to choose between goals that are contradictory or cannot be pursued at the same time. Monocausal explanations, with their emphasis on a single critical factor, underestimate the negative externalities which may be generated on other important factors. In other words, reductionist theories tend to be linear, in the sense that the increases in military effectiveness are supposed to be proportional to increases in the quality or quantity of the factor identified, without anticipating its potential costs in other areas. However, some factors have a nonlinear relationship to military effectiveness, in the sense that they produce diminishing returns beyond an optimal – culminating – point. Although most of the times they are surprising, the world of security frequently experience "trigger effects," provoked by a series of incremental and apparently unimportant events, which ignite a quite disproportionate reaction, like the straw which broke the camel's back. Most, if not all, defense policies involve trade-offs

7 Andreski (1980); Beckley (2010); Brooks (2008); Caverley (2014); Choi (2012); Desch (2008); Lake (1992); Reiter and Stam (2002); Weeks (2014).
8 Delbrück (1990, 296); Kennedy (1987, xvi).
9 Castillo (2014); Posen (1993); Snyder (1991).

and opportunity costs. For instance, seeking security from terrorism involves costs in terms of individual liberty. Seeking total nuclear security through aggressive attempts to induce others to disarm or even launching preventive strikes may result in insecurity or even a nuclear exchange. The prototypical example of a grand strategic trade-off is that between guns and butter, or between the military and civilian economies. Whereas if a state wants to maximize military power in the short term, it should concentrate its expenditures on the military, in the long run, this may damage its civilian economy, that is, the capacity to sustain future defense expenditure. There therefore exists an optimum point to balance expenditure in guns and butter beyond which a further investment would end up being counterproductive.

The third, and perhaps most significant, limit of monocausal explanations consists in the fact that they have a tendency to belittle the strategic dimension, namely, the fact that military effectiveness must be gauged and exercised against a particular opponent for a particular political objective. This means that an effort which can be effective in one context against a particular enemy may not work, for the very same reasons, in another.[10] For example, modern postindustrial societies may be vulnerable to cyberattack, because of their dependence on electronic networks, much more than less advanced societies, which rely on material grids quite immune from cyberwarfare. This means that the various factors need to be gauged not only against one's own characteristics and trade-offs but also against the enemy's. Even the same adversary may change its crucial characteristics within a few years. For example, the French army, which awed its contemporaries by winning in a matter of weeks against the Austrians in 1859, using railways for military purposes for the first time in history, was utterly defeated on the battlefield in two months by Prussia in 1870. A few decades after, that same army proved capable of stopping the German invasion in the Great War, and with British and then American support, it went on to victory. Yet, barely more than twenty years afterward, it was overpowered in six weeks by the German Blitzkrieg. Under pressure from technological development and the rapidly changing mood of public opinion, circumstances today are likely to change even faster and very valuable assessments developed in the decade after the end

[10] Clausewitz (1976, 81).

of the Cold War or even during the Iraqi and Afghan conflicts would now require updating.

Reductionist theories are also inclined to overlook the potential countermeasures that the enemy may take. On this point it is worthy to quote Clausewitz, who famously maintained that "war is not an exercise of the will directed at inanimate matter ... In war the will is directed at an animate object that reacts."[11] The political objectives are actively rebutted by the opponent who is in turn sufficiently sure of his own motives to be prepared to use force, generating "interaction between opposites."[12] This interactive quality of war, which Clausewitz portrays as a "duel on a large scale,"[13] distinguishes strategy from other forms of planning.[14] A plan is the simple scheduling of a series of acts, as a book unfolds from an index of chapters. But in strategy the plans of one side are actively opposed by those of the other: a "good" strategy might paradoxically prove less effective than a "bad" strategy, precisely because it fits in with the enemy's expectations.[15] This is why Field Marshal Von Moltke the Elder cautioned that no plan "survives the first collision with ... the enemy."[16] What tends to carry the day in a violent engagement is the strategy that best foresees and makes provision for the opponent's possible reactions and envisages suitable countermeasures to them.[17] Clausewitz criticizes most military theoreticians for nursing the vain ambition to discover "laws" and "lessons" that are guaranteed to work under all circumstances.[18] Strategy is an art and, as such, can have no fixed rules.[19] A disciple of Clausewitz, the great American strategic thinker Bernard Brodie,[20] reminds us, therefore, that there exist no good and bad strategies but only ones that work "well enough" to achieve the intended results.

Clausewitz also argues that "the very nature of interaction is bound to make it unpredictable."[21] Military force depends on the total quantity of material resources, but also on the will to use them which is "much less easy to determine and can only be gauged approximately by the strength of the motive animating it."[22] As the US discovered in

[11] Clausewitz (1976, 149). [12] Clausewitz (1976, 136).
[13] Clausewitz (1976, 73). [14] Freedman (2013, 16).
[15] Luttwak (1987). [16] Quoted in Rothenberg (1986, 299–300).
[17] Holmes (2007). [18] Clausewitz (1976, 136).
[19] Delbrück (1990 [IV], 293). [20] Brodie (1973, 452).
[21] Clausewitz (1976, 139); cf. also Beyerchen (1992).
[22] Clausewitz (1976, 77).

the Vietnam conflict, it is possible to lose the war even when winning all or most of the battles, if the motivation of the enemy allows it to sustain the conspicuous losses and to refuse capitulation.[23] In the definition of military effectiveness spelled out by Dan Reiter in the introductory chapter, there is an implicit reference to this ultimate trade-off with respect to the "costs" which are deemed "acceptable" in order to accomplish the task at hand. The various measures introduced in pursuit of improved effectiveness must therefore be put in the context of the political objectives which have motivated the use of force in the first place. For example, to fight a conventional power is not the same ball game as facing a nuclear one. To defeat a regular army is a different aspiration than stabilizing a violent society. With good reason some indeed argue that the intensive, firepower oriented approach, currently preferred by Western militaries, is inadequate or even counterproductive in a counterinsurgency situation where, as the typical accounts of the Malayan Emergency remind us, "boots on the ground" may be necessary to win "hearts and minds."

The Other Edge of the Sword

The alternative to the monocausal path is to try and take into account the complex nature of military conflict. In this respect, Clausewitz's notorious conception of the trinity can be an excellent starting point. The Prussian general suggested that the outcome of a war is based on "a remarkable trinity composed of primordial violence, hatred, and enmity which are to be regarded as a blind natural force; of the play of chance and probability within which the creative spirit is free to roam; and of its element of subordination, as an instrument of policy, which makes it subject to reason alone."[24] All three of these factors are "moral" and do not refer to material, quantitative, or technological measures only, quite understandably for Clausewitz's times. Two of the factors also refer to nonmilitary parties, such as the population (which needs to be incited to animosity) and the political leaders, who must be able to choose the final objectives of the war by the rationality of "pure intellect." The intermediate factor related to "chance and probability" is typically military and has to do with the special kinds of "danger, exertion, uncertainty and chance" that arise on the

[23] Mack (1975). [24] Clausewitz (1976, 89).

battlefield.[25] Above all, Clausewitz suggests that no *single* factor is sufficient and that *all* three are necessary.

It has become commonplace to associate the characteristics of the trinity with the main parts of a nation-state: the emotions of the population, the capability to creatively manage "chance and probability" of the military and the "policy" of the government. Since today war among nation-states is claimed to be out of date and most wars are of the civil kind, Clausewitz's thinking has come in for criticism. For example, the changing nature of war, thinks Martin Van Creveld, makes the notion of trinitarian war obsolete: there is no longer a *people* supporting the insurgents, but less clearly defined groups and subsets.[26] Still more recently Sir Rupert Smith has claimed that "war as we have known it" no longer exists as wars "between peoples" have been replaced by "wars amongst the people."[27] It is one of the main arguments of this conclusion that Clausewitz's views are instead still applicable to the new paradigms of war: his trinity continues to be relevant although the key political entities may have changed. For any successful political use of force continues to need the three constituent factors that Clausewitz identified: motivation of the individuals performing military operations, their leaders' skill in guiding them even amid the uneasiness of combat, and the ability of the political leaders to suit the means to the end and choose objectives appropriate to the conflict in progress. These features are necessary, whatever the nature of the political communities involved in a violent contest – be they modern nation-states or less formal and institutionalized organizations – much as in the premodern era when the political entities were disparate and often incoherent.

It is not by coincidence that the three elements of the trinity correspond in terms of domain and scope to the three trade-offs spelled out in the introductory chapter: political support, security threats, and war fighting. Political support has to do with the emotions of the population, security threats matter for government policy and war fighting specifically concerns the military. Some of the chapters deal with just one trade-off, and some with more, while the piece by Jeffrey Friedman argues that cognitive limits, preference for hermeneutic approaches, and self-interest – both as way to manipulate analysis to one's personal or bureaucratic advantage and for blame avoidance

[25] Clausewitz (1976, 104). [26] Van Creveld (1991b). [27] Smith (2005, 1).

later – systematically induce national security officials to forgo precise probabilistic assessments. This is illogical, since "there is little credible evidence suggesting that expressing probability assessments precisely is unwise or infeasible" as well as counterproductive. However, imprecise assessments as to the likelihood of success tend to be inaccurately understood by decision makers, who tend to be overly optimistic. For example, Friedman cites the Bay of Pigs episode, in which the military had estimated a three out of ten probability of success, which it then defined as a "fair chance." President Kennedy then interpreted this assessment as sufficient to proceed, whereas a precise numerical information may have counseled otherwise.

The first trade-off, involving the potential erosion of political support, is explored by a variety of chapters showing that military actions which are too expensive in terms of "blood and treasure" may work on the battlefield but could ultimately undermine the war effort. In particular, Emanuele Castelli and Lorenzo Zambernardi argue that rising concerns for casualties have stimulated the adoption of "force protection" measures. Especially in irregular warfare, force protection may lead to collateral civilian casualties among the populace, whose protection appears as a key factor in the final outcome of the intervention. Likewise, Michael Horowitz examines the potential danger of "blowback" in the use of drones due to their potential for collateral damage, even if he argues that their characteristically lower speed (and expendability) should make them more, rather than less, accurate than their manned counterparts. Rosella Cappella Zielinski concentrates on war financing instead. She maintains that the various ways to finance the war effort are not indifferent to the overall strategy. In particular, a state can rely on direct taxation, which may reduce support for the war, resort to debt financing, which is costly over the longer term, or ask for outside support, thereby limiting its own autonomy since the creditor may influence the war outcome. For instance, when Israel needed American financial support in the 1973 war, it was then forced to accept an early cease-fire against its own wishes.

The second trade-off, involving the possibility that certain efforts may spur new security threats, mainly concerns the grand strategic level. As Dan Reiter reminds us in the introductory chapter, the meaning of victory on the battlefield must be accommodated within overall policy. It is useless "to win" if victory diminishes the overall security of the state. This means that the cost in blood and treasure must be acceptable, and that the conflict must not exacerbate other threats or erode a state's network of diplomatic alliances. One classic illustration of this dynamic is the balance of power mechanism, which facilitates the triggering of counterbalancing alliances if a state accumulates excessive military power, meaning that there is an optimal level of power to guarantee security, beyond which there are diminishing returns. This is only one of the "systemic effects" which bring about outcomes which no actor intended.[28] Focusing on bilateral relations between nuclear powers, Caitlin Talmadge argues that a similar paradoxical mechanism may be at work. If one side is "too" effective and inches closer to a conventional victory, the weaker side may resort to nuclear escalation thereby inflicting unacceptable costs on the other, ultimately frustrating the motivation to go to war in the first place. The losing side may either be concerned with its own territorial survival, in which case it may wish to use nuclear weapons to stop the adversary, or with the survival of its nuclear arsenal. In this case a particularly dangerous situation may ensue, in which the threatened state may have an incentive to use its nuclear weapons before it loses them and the stronger side may accelerate in order to disarm it. This is the reason why vulnerable nuclear powers (such as Israel or North Korea) or conventionally weaker powers (such as NATO versus the Warsaw Pact, Russia versus NATO, or Pakistan versus India) are not inclined to rule out nuclear first use on their part. Castelli and Zambernardi, and Horowitz, speculate on new security threats when they suggest that by inflicting civilian casualties, counterinsurgents may contribute to the insurgent's cause by facilitating its recruitment. Collateral damage may also modify diplomatic alignments by creating an unfavorable opinion in the international community. Cappella Zielinski suggests that excessive expenditure and financial constraints in one conflict may jeopardize the overall position of a state, because it would be forced to cannibalize military commitments elsewhere to sustain the urgent war effort, as has happened to the US in the Korean War. Especially in a multipolar international system like the contemporary one, it may prove dangerous to devote excessive resources to a single conflict.[29]

Concerning the third set of – war-fighting – trade-offs, the book shows how certain tactical choices may be conducive to one purpose, but could undermine other objectives. Force protection and the

[28] Jervis (1997, 132).
[29] Deutsch and Singer (1964); cf. also Simmel (1902, 178, 355).

growing tendency to avoid friendly casualties constrain military options and could encourage ineffective tactics for destroying insurgents, such as a disproportionate use of airpower, limited engagements, and mechanization. Blocking detachments, as argued by Jason Lyall, may enhance the staying power of an army at the expense of increased casualties, less flexible tactical postures, and the risk of mounting grievances among the various army units. Nevertheless, as Lyall explains, blocking detachments were useful in Soviet military efforts at the key battles of Stalingrad and Kursk. This is the reason why they have been adopted by one belligerent out of five in 250 conventional wars between 1800 and 2011. On the other side of the spectrum, relying on professionally and meticulously drilled special forces enhances the prowess of individual units, as Joseph Felter argues with impressive evidence from the Philippines case. However, choosing this path implies putting a limit on the number of troops, due to the costs of training and equipping them. Drones may fulfill their task at a fraction of the human and financial cost of their manned equivalent, and this is why they are proliferating rapidly, but with the risk of hacking and at the expense of mobility and firepower, which are still far superior in traditional aircraft. The introduction of more autonomous robots could in theory solve some of these issues, but it would raise even more disturbing ones, for example in case of programming errors. Finally, aggressive conventional tactics against a nuclear power, such as deep armored penetrations and preemptive aerial strikes against an enemy's infrastructure, may bring about rapid territorial gains at the risk of inducing the enemy, with its back against the wall, to consider escalation.

The three trade-offs indicate that there is no single path to military effectiveness, and different societies may follow successful strategies in disparate ways, depending upon their own characteristics, and those of their adversaries. Although maneuver warfare and Napoleonic decisiveness may seem more spectacular, at times attrition or "Fabian" strategies may prove to be more adequate for the circumstances at hand. Not surprisingly, the eminent military historian Hans Delbrück found merits in both approaches.[30] More recently, Lawrence Freedman has argued that there have always been two distinct types of strategy, glimpsed as early as the *Iliad*: the direct mode counting on physical strength to annihilate the opponent (Achilles's *bíê*), and the indirect

[30] Delbrück (1990, IV: 293–315).

mode where cunning and foxiness are used to manipulate the enemy's expectations and desire for combat (Ulysses's *mêtis*).[31] Sometimes it will be effective to follow technological advances up to the point at which it is conceivable to imagine a robotic "war without warriors" while at other times and for other groups it could be effective to rely on the very opposite solution of "human bombs."[32] Sometimes efficiency and cost cutting may seem the rational choice, but at other times, especially when matters of survival are at stake, a certain redundancy may be better suited, as the prudent duplication (or triplication) of essential security mechanisms on modern airplanes implies.[33]

In sum, all of the chapters in this volume are firmly rooted in the Clausewitzian view of war, confirming the limits of reductionist explanations. Theories about military effectiveness should therefore consider a multicausal avenue to identify regularities, even if its intrinsic complexity favors middle-range theorizing and probabilistic – rather than deterministic – conclusions. Despite today's technological sophistication, military effectiveness still relies, as in the past, on a number of causes, among them a set of factors that may be military or social, material or immaterial. The trade-offs identified in this volume highlight the fact that victory is not a matter of maximizing a single factor but rather resembles a balancing act. Although planning and procurement are essential elements in this endeavor, there is no substitute for the initiative and intuition of the individuals involved at the decisive moment, or for their ability to flexibly adapt to the contingency at hand. Above all, military effectiveness ultimately depends on the wisdom of political leaders, who have to select the "goals" to be assigned to militaries and to be reached by the use of force against the enemy's resolve. Even if, for Clausewitz, war is subject to policy, policy makers must know the boundaries of what the use of force can accomplish.[34] No amount of tactical or operational effectiveness can compensate for a flawed strategy. On this point, circumstances in the twenty-first century have not changed from those of earlier times. As Michael Howard pointed out, "wars still resemble each other more than they resemble any other human activity . . . In all, large bodies of men are trying to impose their will on one another by violence; and in all, events occur which are inconceivable in any other field of experience."[35]

[31] Freedman (2013, 19, 130). [32] Hassner (2007). [33] Landau (1969).
[34] Clausewitz (1976, 87). [35] Howard (1983, 214–215).

References

Andreski, S. 1980. On the Peaceful Disposition of Military Dictatorships. *Journal of Strategic Studies* 3(3): 3–10.

Beckley, M. 2010. Economic Development and Military Effectiveness. *Journal of Strategic Studies* 33(1): 43–79.

Berlin, I. 1953. *The Hedgehog and the Fox*. New York: Simon and Schuster.

Beyerchen, A. 1992. Clausewitz, Nonlinearity and the Unpredictability of War. *International Security* 17(3): 59–90.

Biddle, S. 2004. *Military Power: Explaining Victory and Defeat in Modern Battle*. Princeton, NJ: Princeton University Press.

Brodie, B. 1973. *War and Politics*. New York: Macmillan.

Brooks, R. 2008. *Shaping Strategy: The Civil-Military Politics of Strategic Assessment*. Princeton, NJ: Princeton University Press.

Brooks, R., and E. Stanley, eds. 2007. *Creating Military Power: The Sources of Military Effectiveness*. Stanford, CA: Stanford University Press.

Castillo, J. 2014. *Endurance and War: The National Sources of Military Cohesion*. Stanford, CA: Stanford University Press.

Caverley, J. 2014. *Democratic Militarism: Voting, Wealth, and War*. Cambridge: Cambridge University Press.

Choi, A. 2012. Fighting to the Finish: Democracy and Commitment in Coalition War. *Security Studies* 21(4): 624–653.

Clausewitz, C., von. 1976 (1831). *On War*. Ed. M. Howard and P. Paret. Princeton, NJ: Princeton University Press.

Cohen, E. 2002. *Supreme Command: Soldiers, Statesmen, and Leadership in Wartime*. New York: Free Press.

Delbrück, H. 1990 (1923).*History of the Art of War*. 4 vols. Lincoln: University of Nebraska Press.

Desch, M. 2008. *Power and Military Effectiveness: The Fallacy of Democratic Triumphalism*. Baltimore: Johns Hopkins University Press.

Deutsch, K., and J. D. Singer. 1964. Multipolar Power Systems and International Stability. *World Politics* 16(3): 390–406.

Freedman, L. 2013. *Strategy: A History*. Oxford: Oxford University Press.

Hassner, P. 2007. The Fate of a Century. *The American Interest* 2(6): 36–47.

Holmes, T. M. 2007. Planning versus Chaos in Clausewitz's on War. *Journal of Strategic Studies* 30(1): 129–151.

Horowitz, M. 2010. *The Diffusion of Military Power: Causes and Consequences for International Politics*. Princeton, NJ: Princeton University Press.

Howard, M. 1983. *The Causes of War and Other Essays*. Cambridge, MA: Harvard University Press.

Jervis, R. 1997. *System Effects: Complexity in Political and Social Life*. Princeton, NJ: Princeton University Press.

Johnston, A. 1995. Thinking about Strategic Culture. *International Security* 19(4): 32–64.

Kennedy, P. 1987. *The Rise and Fall of the Great Powers*. New York: Random House.

Knox, M., and W. Murray. 2001. *The Dynamics of Military Revolution*. Cambridge: Cambridge University Press.

Krepinevich, A. 1994. Cavalry to Computer: The Pattern of Military Revolutions. *The National Interest* 37(Fall): 30–42.

Lake, D. 1992. Powerful Pacifists: Democratic States and War. *American Political Science Review* 86(1): 24–37.

Landau, M. 1969. Redundancy, Rationality, and the Problem of Duplication and Overlap. *Public Administration Review* 29(4): 346–358.

Luttwak, E. 1987 *Strategy: The Logic of War and Peace*. Cambridge, MA: Belknap Press.

Mack, A. 1975. Why Big Nations Lose Small Wars: The Politics of Asymmetric Conflict. *World Politics* 27(2): 175–200.

Pape, R. 1996. *Bombing to Win*. Ithaca, NY: Cornell University Press.

Posen, B. 1993. Nationalism, the Mass Army and Military Power. *International Security* 18(2): 80–124.

Reiter, D., and Allan C. Stam. 2002. *Democracies at War*. Princeton, NJ: Princeton University Press.

Rosen, S. 1991. *Winning the Next War: Innovation and the Modern Military*. Ithaca, NY: Cornell University Press.

Rothenberg, G. 1986. Moltke, Schlieffen and the Doctrine of Strategic Envelopment. In Peter Paret, ed., *Makers of Modern Strategy from Machiavelli to the Nuclear Age*, 296–325. Princeton, NJ: Princeton University Press.

Shils, E., and M. Janowitz. 1948. Cohesion and Disintegration in the Wehrmacht in World War II. *Public Opinion Quarterly* (12)2: 280–315.

Simmel, G. 1902. The Number of Members as Determining the Sociological Form of the Group (Part II). *American Journal of Sociology* 8(2): 158–196.

Smith, R. 2005. *The Utility of Force: The Art of War in the Modern World*. New York: Knopf.

Snyder, J. 1984. *Ideology of the Offensive: Military Decision Making and the Disasters of 1914*. Ithaca, NY: Cornell University Press.

1991. *Myths of Empire: Domestic Politics and International Ambition*. Ithaca, NY: Cornell University Press.

Stam, Allan C. 1996. *Win, Lose, or Draw: Domestic Politics and the Crucible of War*. Ann Arbor: University of Michigan Press.

Talmadge, C. 2015. *The Dictator's Army: Battlefield Effectiveness in Authoritarian Regimes*, Ithaca, NY: Cornell University Press.

Van Creveld, M. 1991a. *The Transformation of War*. New York: Free Press.

1991b. *Technology and War*. New York: Simon and Schuster.

Weeks, J. 2014. *Dictators at War and Peace*. Ithaca, NY: Cornell University Press.

Index